A Photographic Atlas
for
Anatomy and Physiology

BRIEF EDITION

John L. Crawley

Kent M. Van De Graaff
Weber State University

Morton Publishing Company
925 W. Kenyon, Unit 12
Englewood, Colorado 80110

www.morton-pub.com

To our families for their sacrifice, support, and love.

Copyright 2002 Morton Publishing Company

ISBN: 0-89582-609-7

10 9 8 7 6 5 4 3 2 1

Printed in the United States of America

Table of Contents

A Photographic Atlas for Anatomy and Physiology Brief Edition

Preface

Human anatomy is the scientific discipline that investigates the structure of the body, and human physiology is the scientific discip that investigates how body structures function. These subjects may be taught independent of each other in separate courses, or they be taught together in integrated anatomy and physiology courses. Regardless of whether or not anatomy is taught independently f physiology or if the two disciplines are integrated as a single course, it is necessary for a student to have a conceptualized visualizatio body structure and a knowledge of its basic descriptive anatomical terminology in order to understand how the body functi Knowledge of the structure and function of the body is an essential foundation for each of the allied-health disciplines.

A Photographic Atlas for the Anatomy and Physiology Brief Edition is designed for all students taking separate or integra courses in human anatomy and physiology. This full-color atlas can accompany and will augment any human anatomy, hur physiology, or combined human anatomy and physiology textbook. It is designed to be of particular value to students in a labora situation and could either accompany a laboratory manual or, in certain courses, serve as the laboratory manual. Health professio who benefit from a knowledge of the structure and function of the body include: nurses, emergency medical technicians, radiolog assistants, respiratory therapists, massage therapists, sports medicine therapists, nursing assistants, physician assistants, and phys therapists. This atlas is prepared to meet the needs of students in these and other allied-health fields.

Anatomy and physiology are visually oriented sciences. Great care has gone into the preparation of this photographic atlas to pro students with a complete set of photographs for each of the human body systems. Human cadavers have been carefully dissected photographs taken to clearly depict each of the principal organs from each of the body systems.

A visual balance is achieved in this atlas between the various levels available to observe the structure of the body. Microsco anatomy is presented by photomicrographs at the light microscope level and electron micrographs from scanning and transmiss electron microscopy. Carefully selected photographs are used throughout the atlas to provide a balanced perspective of the g anatomy. Great care has been taken to construct completely labeled informative figures that are depicted clearly and accurately. terminology used in this atlas are those that are approved and recommended by the Basle Nomina Anatomica (BNA). The s terminology is used in most of the current human anatomy and physiology texts.

Advise for Studying Human Anatomy and Physiology

A Photographic Atlas for the Anatomy and Physiology Brief Edition was written to help you achieve competency in human anato and physiology. An understanding of the structure and function of the human body will provide you with a solid foundation for m health or fitness-related disciplines. You will be exposed to a vast amount of new information during this course; therefore assimilate and retain the information presented in this laboratory atlas, you might follow these suggestions:

1. Prepare in advance. Examine carefully each chapter before attending the laboratory period. Come to the laboratory ready to m observations and relate the new information to facts previously learned. As you are introduced to each new term or anatomical struc in your studying, identify each label where it is located on the accompanying illustrations. Anatomy is a very descriptive scie and many of the terms are of Greek or Latin origin. Refer to a glossary in your textbook for help in learning the descriptive root these words. If you understand the descriptive roots of these words, each term will have more meaning and will perhaps be ea to remember.

Make notes to yourself of the things you should or would like to observe during the laboratory period. The most profitable productive time of this course should be during the laboratory period and it will be only if you are sufficiently prepared in adva

2. Study on a regular basis. Keeping up with the daily addition of material is imperative for the retention of the mate Establishing a set time each day to study is an integral part of developing good study habits. Reviewing and self-testing after e laboratory exercise are excellent ways to reinforce your retention. It is frequently helpful to write a term several times so a establish a mental impression of that word and its meaning.

3. Review with another person. This technique for studying can be beneficial in several ways. It not only sets aside a specific t for studying but also helps with the pronunciation of new terms. Retention improves when a term is repeatedly spoken. Studying v another person helps you identify weaknesses or inconsistencies in your notes. This type of studying is most effective, however, a you have spent a considerable length of time studying independently.

4. Make anatomy relevant. Use yourself as a model from which to learn. As you learn a bone or a process on a bone, palpate part of your body. Contract the muscles you are studying so you can better understand their actions. Listen with a stethoscope to sounds of the heart as you are learning the location and functions of the valves. If you do this along with along with the o suggestions offered, anatomy and physiology will be easier to learn and will be one of the most exciting courses you will ever h an opportunity to take.

Body Organization

Anatomy is the study of body structures. An example of an anatomical study is learning about the structure of the heart—the chambers, valves, and vessels that serve the heart muscle. *Physiology* is the study of body function. An example of a physiological study is learning what causes the heart muscles to contract—the sequence of blood flow through the heart and what causes blood pressure. The anatomy (structure) and the physiology (function) of any part of the body are always related, or in other words, structure determines function.

Most of the physiological processes within the body act to maintain *homeostasis*. Simply defined, homeostasis is maintaining near consistent internal conditions within the body despite changing conditions in the external environment. For example, one area of your brain acts as a thermostat to keep your body temperature near 37°C (98.6°F). Being too warm causes you to sweat and cool the body, while being cold causes you to shiver and warm the body. Maintaining overall body homeostasis is achieved through many interacting physiological processes involving all levels of body organization, and is absolutely necessary for survival.

Structural and functional levels of organization exist in the body, and each of its parts contributes to the total organism. In the study of human anatomy and physiology, six levels of body organization are generally recognized—the molecular level, the cellular level, the tissue level, the organ level, the systems level, and the organismic level (fig. 1.1).

Cells are microscopic and are the smallest living part of all organisms. *Tissues* are layers of groups of similar cells that perform specific functions. An *organ* is an aggregate of two or more tissues integrated to perform a particular function. The *systems* of the body consist of various body organs that have similar or related functions. All the systems of the body are interrelated and function together constituting the *organism*.

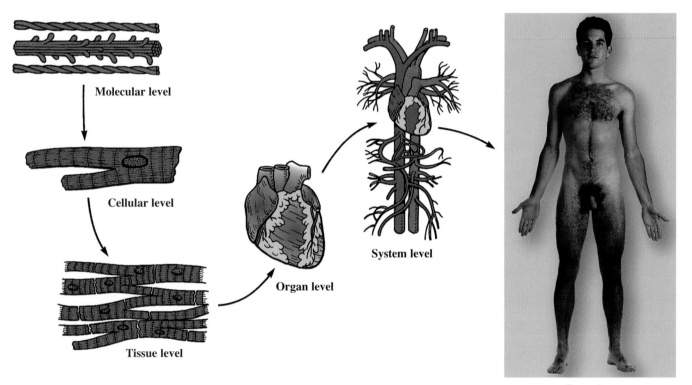

Molecular level

Cellular level

Tissue level

Organ level

System level

Organism

Figure 1.1 The levels of structural organization and complexity within the body.

Table 1.1 Directional Terminology for Describing Human Body Structures

Term	Definition	Example
Superior (cephalic, cranial)	Toward the top of the head	The neck is superior to the thorax.
Inferior (caudal)	Away from the top of the head	The pubic region is inferior to the abdomen.
Anterior (ventral)	Toward the front of the body	The eyes, nose, and mouth are on the anterior side of the body.
Posterior (dorsal)	Toward the back of the body	The spinal cord extends down the posterior side of the body.
Lateral	Toward the side of the body	The arms are on the lateral sides of the body.
Medial	Toward the median plane of the body	The heart is medial to the lungs.
Superficial (external)	Toward the surface of the body	The skin is superficial to the muscles.
Deep (internal)	Away from the surface of the body	The heart is positioned deep within the thoracic cavity.
Parietal	Reference to the body wall of the trunk (thorax and abdomen)	The parietal peritoneum is the membrane lining the abdominal cavity.
Visceral	Reference to internal organs of trunk	The stomach is covered by a thin membrane called the visceral peritoneum.

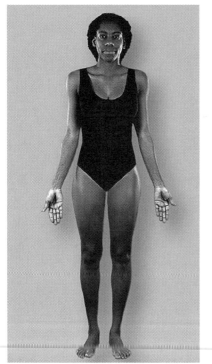

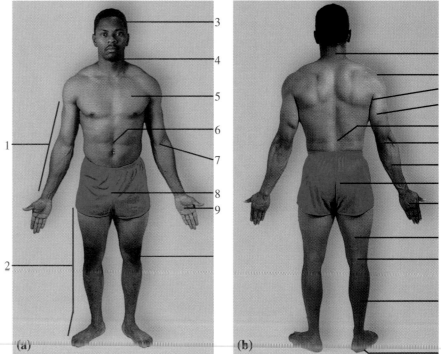

Figure 1.2 The anatomical position provides a basis of reference for describing the relationship of one body part to another. In the anatomical position, the person is standing, the feet are parallel, the eyes are directed forward, and the arms are to the sides with the palms turned forward and the fingers are pointed straight down.

Figure 1.3 Major body parts and regions in humans (bipedal vertebrate). (a) An anterior view and (b) a posterior view.

1. Upper extremity
2. Lower extremity
3. Head
4. Neck, anterior aspect
5. Thorax (chest)
6. Abdomen
7. Cubital fossa
8. Pubic region

9. Palmar region (palm)
10. Patellar region (patella)
11. Cervical region
12. Shoulder
13. Axilla (armpit)
14. Brachium (upper arm)
15. Lumbar region
16. Elbow

17. Antebrachium (forearm)
18. Gluteal region (buttock)
19. Dorsum of hand
20. Thigh
21. Popliteal fossa
22. Calf
23. Plantar surface (sole)

Figure 1.4 Directional terminology and superficial structures in a fetal pig (quadrupedal vertebrate).

1. Anus
2. Tail
3. Scrotum
4. Knee
5. Tcat
6. Ankle
7. Umbilical cord
8. Hoof
9. Auricle (pinna)
10. External auditory canal
11. Superior palpebra (superior eyelid)
12. Elbow
13. Wrist
14. Naris (nostril)
15. Tongue

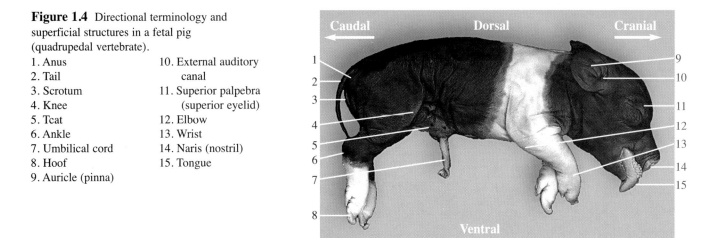

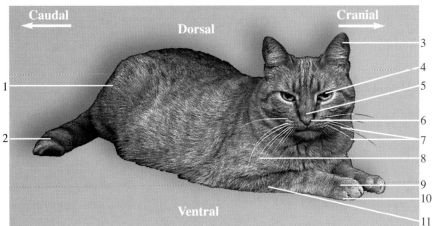

Figure 1.5 Directional terminology and superficial structures in a cat (quadrupedal vertebrate).

1. Thigh
2. Tail
3. Auricle (pinna)
4. Superior palpebra (superior eyelid)
5. Bridge of nose
6. Naris (nostril)
7. Vibrissae
8. Brachium
9. Manus (front foot)
10. Claw
11. Antebrachium

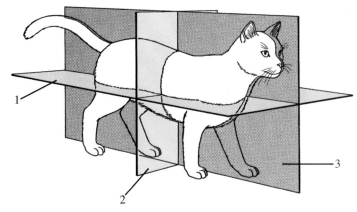

Figure 1.6 Planes of reference in a cat.
1. Coronal plane (frontal plane)
2. Transverse plane (cross-sectional plane)
3. Sagittal plane

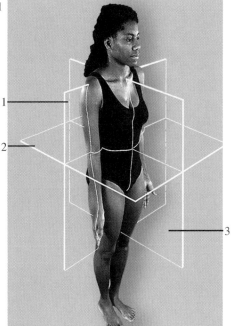

Figure 1.7 Planes of reference in a human (bipedal vertebrate).
1. Coronal plane (frontal plane)
2. Transverse plane (cross-sectional plane)
3. Sagittal plane

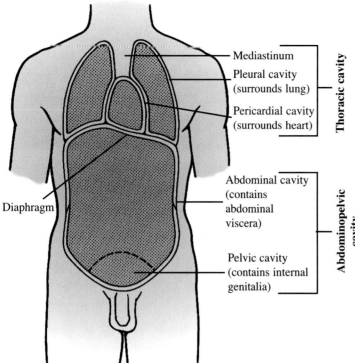

Figure 1.8 An anterior view of the body cavities of the trunk.

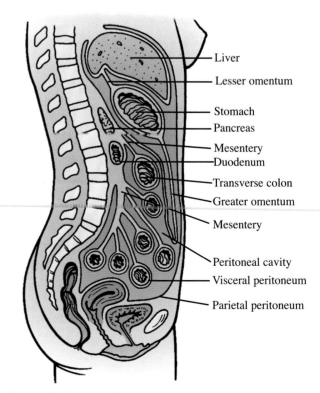

Figure 1.9 MR image of the trunk showing the body cavities and their contents.

1. Thoracic cavity
2. Abdominopelvic cavity
3. Image of heart
4. Image of diaphragm
5. Image of rib
6. Image of lumbar vertebra
7. Image of Ilium

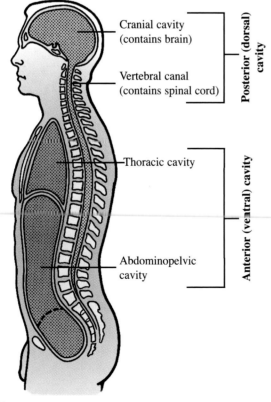

Figure 1.10 A midsagittal view of the body cavities.

Figure 1.11 A midsagittal view of the organs of the abdominopelvic cavity and their supporting membranes.

Figure 1.12 Human male.
(a) An anterior view and
(b) a posterior view.
  1. Facial region
  2. Cranial region
  3. Posterior neck
  4. Anterior neck
  5. Shoulder
  6. Thorax
  7. Nipple
  8. Brachium
  9. Elbow
  10. Cubital fossa
  11. Abdomen
  12. Umbilicus (navel)
  13. Antebrachium
  14. Wrist
  15. Hand
  16. Natal (gluteal) cleft
  17. Fold of buttock
  18. External genitalia
  19. Thigh
  20. Patella
  21. Popliteal fossa
  22. Leg
  23. Ankle
  24. Foot

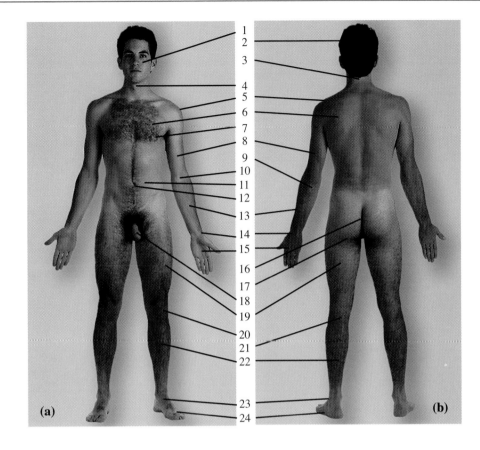

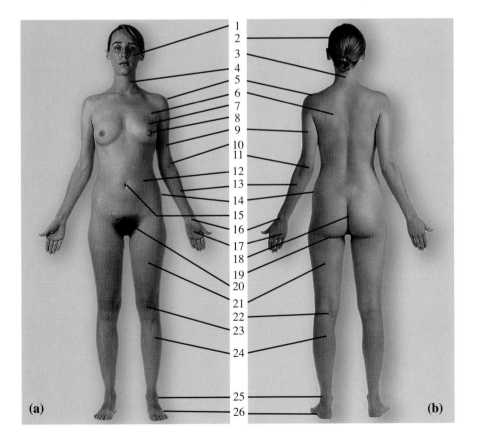

Figure 1.13 Human
female. (a) An anterior view
and (b) a posterior view.
  1. Facial region
  2. Cranial region
  3. Posterior neck
  4. Anterior neck
  5. Shoulder
  6. Thorax
  7. Breast
  8. Nipple
  9. Brachium
  10. Cubital fossa
  11. Elbow
  12. Abdomen
  13. Antebrachium
  14. Iliac crest
  15. Umbilicus (navel)
  16. Wrist
  17. Hand
  18. Natal (gluteal) cleft
  19. Fold of buttock
  20. Mons pubis
  21. Thigh
  22. Popliteal fossa
  23. Patella
  24. Leg
  25. Ankle
  26. Foot

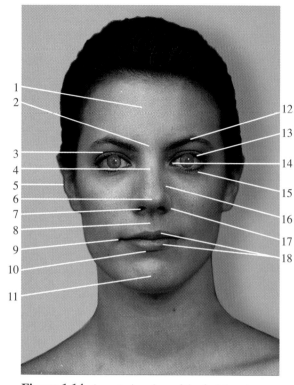

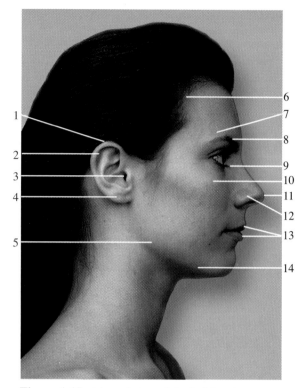

Figure 1.14 An anterior view of the facial region.

1. Forehead
2. Root of nose
3. Superior palpebral sulcus
4. Bridge of nose
5. Auricle (pinna)
6. Apex of nose
7. Nostril
8. Philtrum
9. Corner of mouth
10. Mentolabial sulcus
11. Chin (mentalis)
12. Eyebrow
13. Eyelashes of upper eyelid
14. Lacrimal caruncle
15. Eyelashes of lower eyelid
16. Nasofacial angle
17. Alar nasal sulcus
18. Lips

Figure 1.15 A lateral view of the facial region.

1. Helix of auricle
2. Antihelix
3. External auditory canal
4. Earlobe
5. Angle of mandible
6. Hair line
7. Superciliary ridge
8. Eyebrow
9. Eyelashes
10. Zygomatic arch
11. Apex of nose
12. Ala nasi
13. Lips
14. Body of mandible

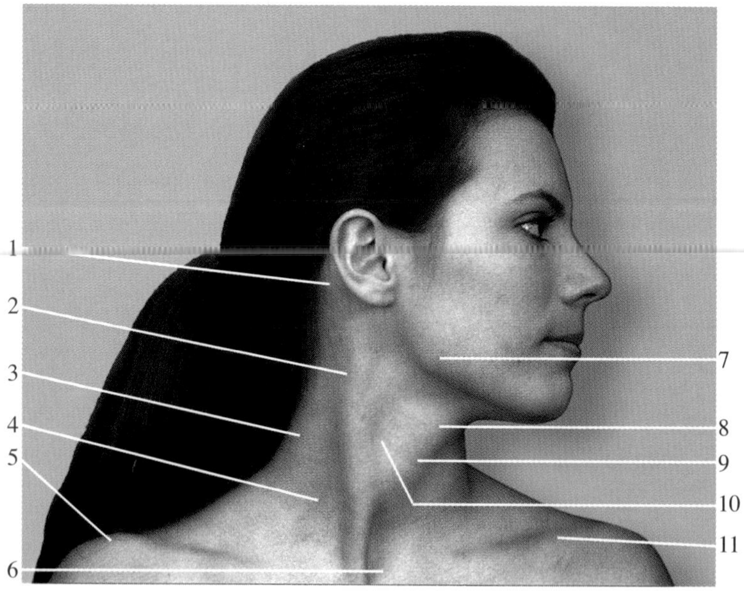

Figure 1.16 An anterolateral view of the neck. (m. = muscle)

1. Mastoid process
2. Sternocleidomastoid m.
3. Trapezius m.
4. Posterior triangle of neck
5. Acromion of scapula
6. Jugular notch
7. Angle of mandible
8. Hyoid bone
9. Thyroid cartilage of larynx
10. Anterior triangle of neck
11. Clavicle

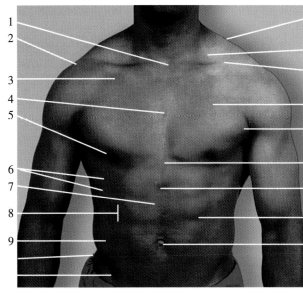

Figure 1.17 An anterior view of the thorax and abdomen.

1. Jugular notch
2. Acromion of scapula
3. Clavipectoral triangle
4. Sternum
5. Nipple
6. Serratus anterior m.
7. Rectus abdominis m.
8. Linea semilunaris
9. External abdominal oblique m.
10. Iliac crest
11. Inguinal ligament
12. Trapezius m.
13. Supraclavicular fossa
14. Clavicle
15. Pectoralis major m.
16. Anterior axillary fold
17. Xiphoid process
18. Linea alba
19. Tendinous inscription
20. Umbilicus

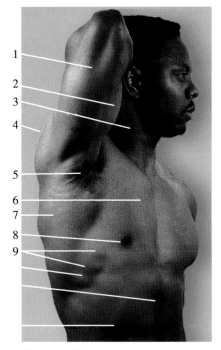

Figure 1.18 An anterolateral view of the thorax, abdomen, and axilla.

1. Triceps brachii m.
2. Biceps brachii m.
3. Sternocleidomastoid m.
4. Deltoid m.
5. Axilla
6. Pectoralis major m.
7. Latissimus dorsi m.
8. Nipple
9. Serratus anterior m.
10. Intercostal m.
11. Linea alba
12. External abdominal oblique m.

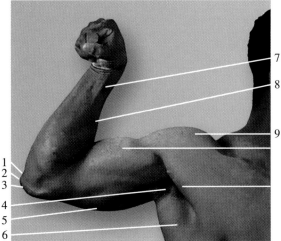

Figure 1.19 Right shoulder, axilla, and upper extremity.

1. Olecranon of ulna
2. Ulnar Sulcus
3. Medial epicondyle of humerus
4. Axilla
5. Triceps brachii m.
6. Latissimus dorsi m. (posterior axillary fold)
7. Tendon of flexor carpi radialis longus m.
8. Brachioradialis m.
9. Deltoid m.
10. Biceps brachii m.
11. Pectoralis major m. (anterior axillary fold)

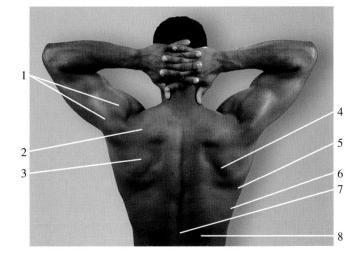

Figure 1.20 A posterior view of the thorax.

1. Deltoid m.
2. Trapezius m.
3. Infraspinatus m.
4. Triangle of auscultation
5. Inferior angle of scapula
6. Latissimus dorsi m.
7. Median furrow over vertebral column
8. Erector spinae m.

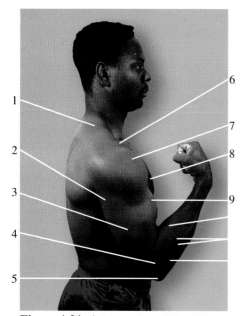

Figure 1.21 A lateral view of the right shoulder and upper extremity. (m. = muscle mm. = muscles)
1. Trapezius m.
2. Long head of triceps brachii m.
3. Lateral head of triceps brachii m.
4. Lateral epicondyle of humerus
5. Olecranon of ulna
6. Acromion of scapula
7. Deltoid m.
8. Pectoralis major m.
9. Biceps brachii m.
10. Brachioradialis m.
11. Extensor carpi radialis longus and brevis mm.
12. Extensor digitorum m.

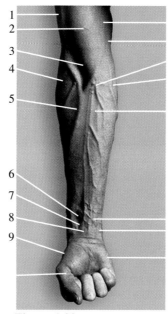

Figure 1.22 An anterior view of the right upper extremity.
1. Cephalic vein
2. Biceps brachii m.
3. Cubital fossa
4. Brachioradialis m.
5. Cephalic vein
6. Site for palpation of radial artery
7. Tendon of flexor carpi radialis m.
8. Tendon of palmaris longus m.
9. Thenar eminence
10. Metacarpophalangeal joint of thumb
11. Site of palpation of brachial artery
12. Basilic vein
15. Median antebrachial vein
16. Tendon of superficial digital flexor m.
17. Styloid process of ulna
18. Hypothenar eminence

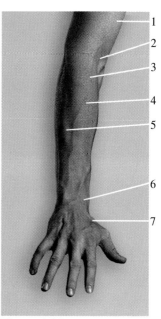

Figure 1.23 A posterior view of the right upper extremity.
1. Biceps brachii m.
2. Cubital fossa
3. Brachioradialis m.
4. Extensor carpi radialis longus m.
5. Extensor carpi ulnaris m.
6. Styloid process of radius
7. Tendon of extensor pollicus longus m.

Figure 1.24 An anterior view of the right hand.
1. Tendon of flexor carpi radialis m.
2. Tendon of palmaris longus m.
3. Flexion creases on wrist
4. Thenar eminence
5. Hypothenar eminence
6. Flexion creases on palm of hand
7. Flexion creases on third digit

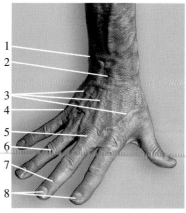

Figure 1.25 A posterior view of the right hand.
1. Styloid process of ulna
2. Position of extensor retinaculum
3. Tendons of extensor digitorum m.
4. Tendon of extensor digiti minimi m.
5. Metacarpophalangeal joint
6. Proximal interphalangeal joint
7. Distal interphalangeal joint
8. Nails

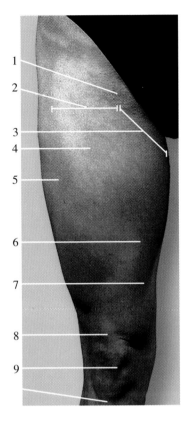

Figure 1.26 An anterior view of the right thigh.
1. Site of femoral triangle
2. Quadriceps femoris group of muscles
3. Adductor group of muscles
4. Rectus femoris m.
5. Vastus lateralis m.
6. Sartorius m.
7. Vastus medialis m.
8. Tendon of quadriceps femoris m.
9. Patella
10. Patellar ligament

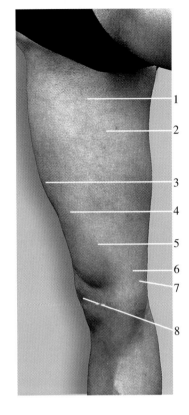

Figure 1.27 A medial view of the right thigh.
1. Adductor magnus m.
2. Gracilis m.
3. Rectus femoris m.
4. Sartorius m.
5. Vastus medialis m.
6. Semimembranosus m.
7. Semitendinosus m.
8. Patella

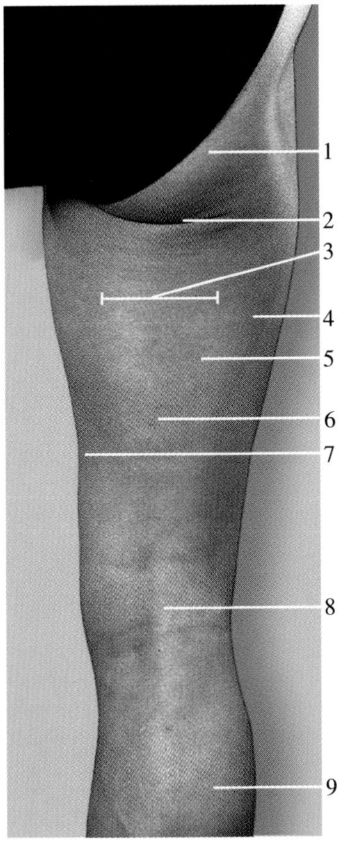

Figure 1.28 A posterior view of the right thigh.
1. Gluteus maximus m.
2. Fold of buttock (gluteal fold)
3. Hamstring group of muscles
4. Vastus lateralis m.
5. Long head of biceps femoris m.
6. Semitendinosus m.
7. Gracilis m.
8. Popliteal fossa
9. Lateral head of gastrocnemius m.

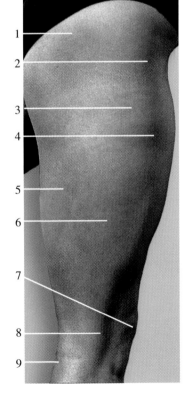

Figure 1.29 A lateral view of the right thigh.
1. Gluteus maximus m.
2. Tensor fasciae latae m.
3. Vastus lateralis m.
4. Rectus femoris m.
5. Biceps femoris m.
6. Iliotibial tract
7. Patella
8. Lateral epicondyle of femur
9. Popliteal fossa

Figure 1.30 An anterior view of the right leg and foot.
1. Patclla
2. Patellar ligament
3. Tibialis anterior m.
4. Lateral malleolus of fibula
5. Medial malleolus of tibia
6. Site for palpation of dorsal pedis artery
7. Tendons of extensor digitorum longus m.
8. Tendon of extensor hallucis longus m.

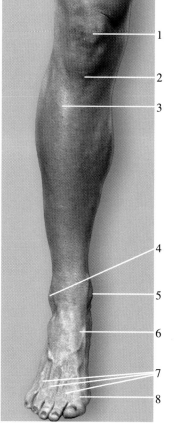

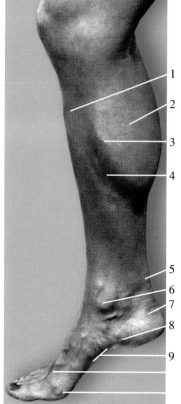

Figure 1.31 A medial view of the right leg and foot.
1. Tibia
2. Medial head of gastrocnemius m.
3. Branch of great saphenous vein
4. Soleus m.
5. Tendo calcaneus
6. Medial malleolus of tibia
7. Calcaneus
8. Abductor hallucis m.
9. Longitudinal arch
10. Tendon of extensor hallucis longus m.
11. Head of first metatarsal bone

Figure 1.32 A posterior view of the right leg and foot.
1. Popliteal fossa
2. Lateral head of gastrocnemius m.
3. Medial head of gastrocnemius m.
4. Soleus m.
5. Peroneus longus m.
6. Tendo calcaneus
7. Peroneus brevis m.
8. Medial malleolus of tibia
9. Lateral melleolus of fibula
10. Calcaneus
11. Abductor digiti minimi m.
12. Plantar surface of foot

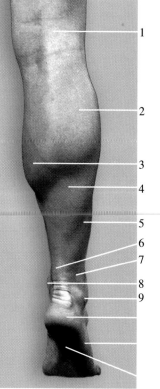

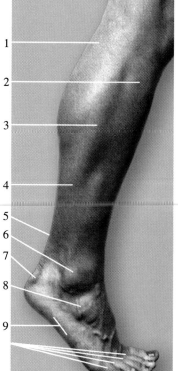

Figure 1.33 A lateral view of the right leg and foot.
1. Lateral head of gastrocnemius m.
2. Tibialis anterior m.
3. Peroneus longus m.
4. Soleus m.
5. Tendo calcaneus
6. Lateral malleolus of fibula
7. Calcaneus
8. Extensor digitorum brevis m.
9. Lateral surface of foot
10. Tendons of extensor digitorum longus m.

Cells

Cells are the basic structural and functional units of organization within the body. Although diverse, human cells have structural similarities including a *nucleus* containing a nucleolus, various *organelles* suspended in *cytoplasm*, and an enclosing *cell (plasma) membrane* (fig. 2.1).

The *nucleus* is the large spheroid body within a cell that contains the *chromatin*, *nucleolus*, and *nucleoplasm*—the genetic material of the cell. The nucleus is enclosed by a double membrane called the *nuclear membrane*, or *nuclear envelope*. The *nucleolus* is a dense, nonmembranous body composed of protein and RNA molecules. The chromatin consists of fibers of protein and DNA molecules. Prior to cellular division, the chromatin shortens and coils into rod-shaped *chromosomes*. Chromosomes consists of DNA and proteins called *histones*.

The *cytoplasm* of a cell is the medium of support between the nuclear membrane and the cell membrane. *Organelles* are minute membrane-bound structures within the cytoplasm of a cell that are concerned with specific functions. The cellular functions carried out by the organelles are referred to as *metabolism*. The principal organelles and their functions are listed in table 2.1. In order for cells to remain alive, metabolize, and maintain homeostasis, certain requirements must be met. These include having access to nutrients and oxygen, being able to eliminate wastes, and being maintained in a constant, protective environment.

The cell membrane is composed of phospholipid and protein molecules, which gives form to a cell and controls the passage of material into and out of a cell. More specifically, the proteins in the cell membrane provide: 1) structural support; 2) a mechanism of molecule transport across the membrane; 3) enzymatic control of chemical reactions; 4) receptors for hormones and other regulatory molecules; and 5) cellular markers (antigens), which identify the blood and tissue type. The carbohydrate molecules: 1) repel negative objects due to their negative charge; 2) act as receptors for hormones and other regulatory molecules; 3) form specific cell markers which enable like cells to attach and aggregate into tissues; and 4) enter into immune reactions.

The permeability of the cell membrane is a function of: 1) size of molecules; 2) solubility in lipids; 3) ionic charge of molecules; and 4) the presence of carrier molecules.

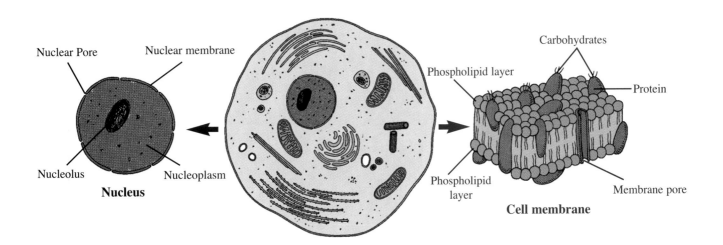

Figure 2.1 A cell and it's nucleus and cell (plasma) membrane.

Table 2.1 Structure and Function of Cellular Components

Component	Structure	Function
Cell (plasma) membrane	Composed of protein and phospholipid molecules	Provides form to cell and controls passage of materials into and out of cell
Cytoplasm	Fluid to jelly-like substance	Suspends organelles and provides a matrix in which chemical reactions occur
Endoplasmic reticulum	Interconnecting hollow channels	Provides supporting framework of cell; facilitates cell transport
Ribosomes	Granules of nucleic acid	Synthesize proteins
Mitochondria	Double-layered sacs with cristae	Production of ATP
Golgi complex	Flattened sacs with vacuoles	Synthesize carbohydrates and packages molecules for secretion
Lysosomes	Membrane-surrounded sacs of enzymes	Digest foreign molecules and worn cells
Centrosome	Mass of two rodlike centrioles	Organizes spindle fibers and assists mitosis
Vacuoles	Membranous sacs	Store and excrete substances within the cytoplasm
Fibrils and microfibrils	Protein strands	Support cytoplasm and transport materials
Cilia and flagella	Cytoplasmic extensions from cell	Movement of particles along cell surface or move cell
Nucleus	Nuclear membrane, nucleolus, and chromatin (DNA)	Directs cell activity; forms ribosomes

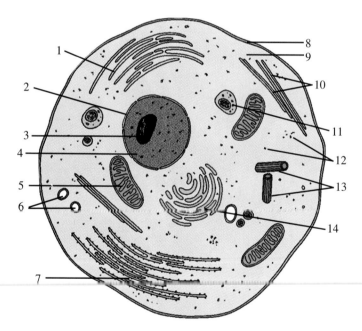

Figure 2.2 A typical animal cell.

1. Smooth endoplasmic reticulum
2. Nuclear membrane
3. Nucleolus
4. Nucleoplasm
5. Mitochondrion
6. Vesicles
7. Rough endoplasmic reticulum

8. Cell membrane
9. Cytoplasm
10. Microtubules
11. Lysosome
12. Ribosomes
13. Centrioles
14. Golgi complex

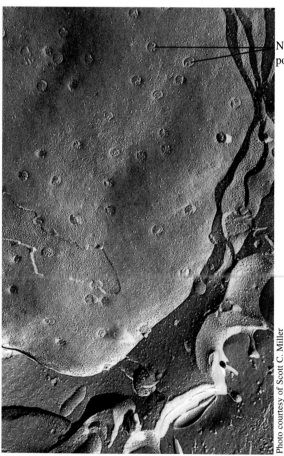

Nuclear pores

Photo courtesy of Scott C. Miller

Figure 2.3 An electron micrograph of a freeze   1000X
fractured nuclear envelope showing the nuclear pores.

Photo courtesy of Scott C. Miller

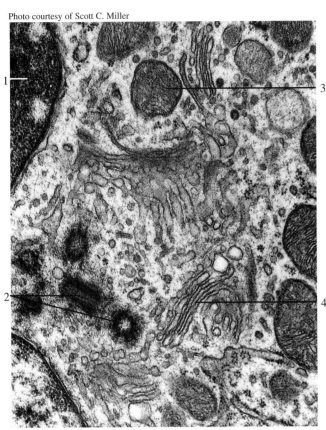

Figure 2.4 An electron micrograph of various organelles.
1. Nucleus   3. Mitochondrion
2. Centrioles   4. Golgi complex

Photo courtesy of Scott C. Miller

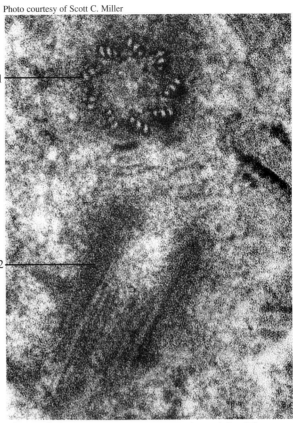

Figure 2.5 An electron micrograph of centrioles. The centrioles are positioned at right angles to one another.
1. Centriole   2. Centriole

Photo courtesy of Scott C. Miller

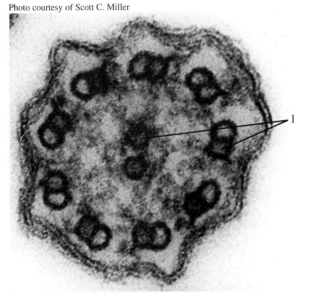

Figure 2.6 An electron micrograph of cilia showing the characteristic "9 + 2" arrangement of microtubules in the cross sections.
1. Microtubules

Photo courtesy of Scott C. Miller

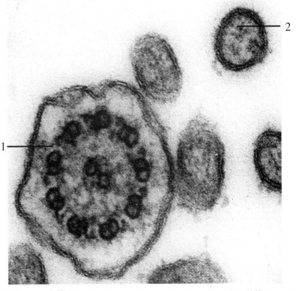

Figure 2.7 An electron micrograph showing the difference between a microvillus and a cilium.
1. Microvillus   2. Cilium

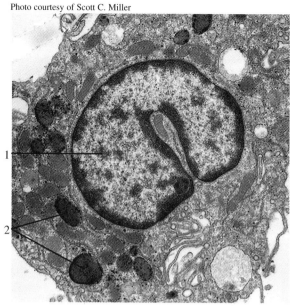

Figure 2.8 An electron micrograph of lysosomes.
1. Nucleus
2. Lysosomes

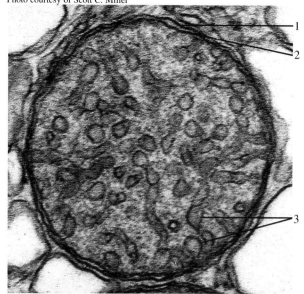

Figure 2.9 An electron micrograph of a mitochondrion.
1. Outer membrane
2. Crista
3. Inner membranes

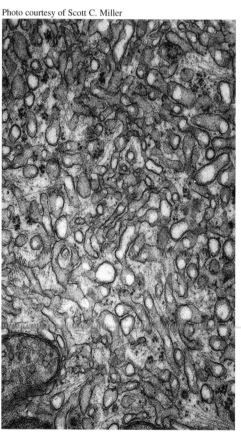

Figure 2.10 An electron micrograph of smooth endoplasmic reticulum from the testis.

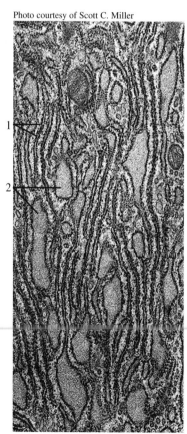

Figure 2.11 An electron micrograph of rough endoplasmic reticulum.
1. Ribosomes   2. Cisternae

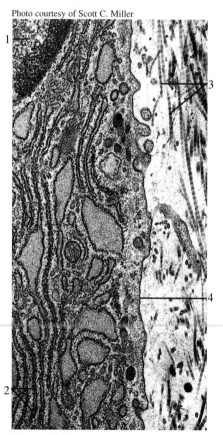

Figure 2.12 Rough endoplasmic reticulum secreting collagenous filaments to the outside of the cell.
1. Nucleus   3. Collagenous
2. Rough endoplasmic   filaments
   reticulum   4. Cell membrane

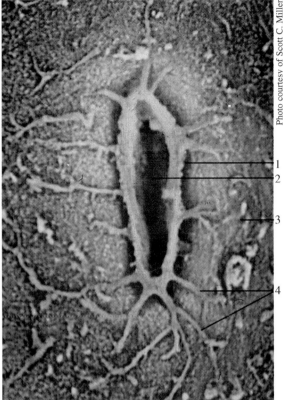

Figure 2.13 Adipocytes in adipose tissue.   200X
1. Cell membrane of adipocyte
2. Lipid-filled vacuole of adipocyte
3. Nucleus

Figure 2.14 An electron micrograph of an osteocyte in cortical bone matrix.
1. Lacuna   3. Bone matrix
2. Osteocyte   4. Canaliculi

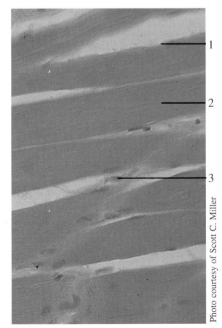

Figure 2.15 Skeletal muscle cells (fibers).
1. Sarcolemma
2. Striations
3. Nucleus

Figure 2.16 An electron micrograph of an erythrocyte (red blood cell).

Photo courtesy of Scott C. Miller

Figure 2.17 An electron micrograph of a skeletal muscle myofibril, showing the striations.

1. Mitochondria
2. Z line
3. I band
4. A band
5. T-tubule
6. Sarcoplasmic reticulum
7. H Band
8. Sacromere

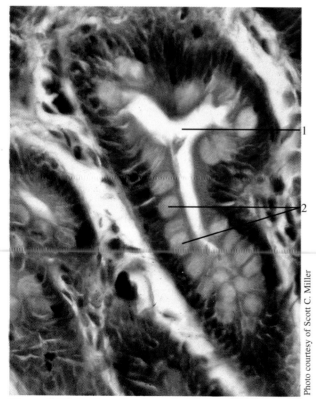

Photo courtesy of Scott C. Miller

Figure 2.18 Goblet cells within an intestinal gland   430X
(crypt of Lieberkühn) of small intestine.

1. Lumen of gland
2. Goblet cells

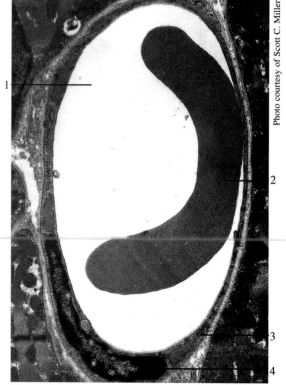

Photo courtesy of Scott C. Miller

Figure 2.19 An electron micrograph of a capillary
containing an erythrocyte.

1. Lumen of capillary
2. Erythrocyte
3. Endothelial cell
4. Nucleus of endothelial cell

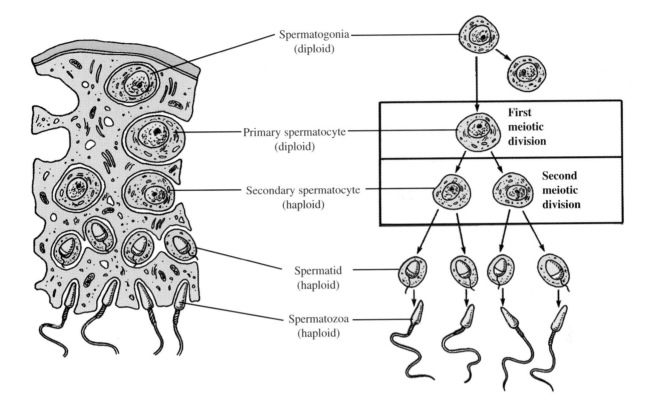

Figure 2.20 Spermatogenesis is the production of male gametes, or spermatozoa, through the process of meiosis.

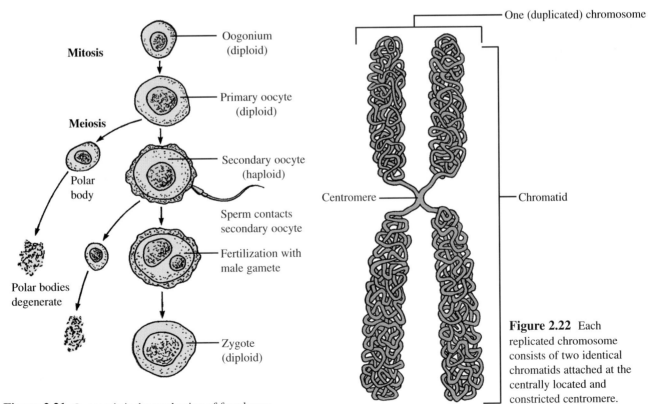

Figure 2.21 Oogenesis is the production of female sex gametes, or ova, through the process of meiosis.

Figure 2.22 Each replicated chromosome consists of two identical chromatids attached at the centrally located and constricted centromere.

Figure 2.23 Stages of mitosis.

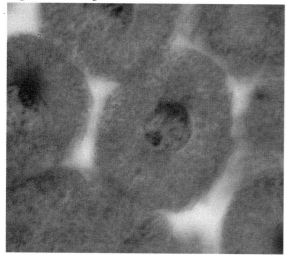

Prophase

Each chromosome consists of two chromatids jointed by a centromere. Spindle fibers extend from each centriole.

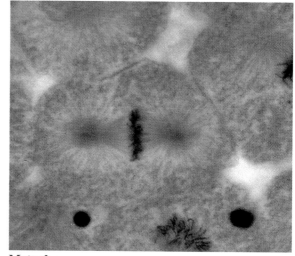

Metaphase

The chromosomes are positioned at the equator. The spindle fibers from each centriole attach to the centromeres.

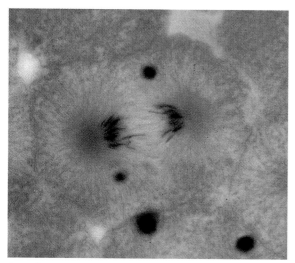

Anaphase

The centromeres split, and the sister chromatids separate as each is pulled to an opposite pole.

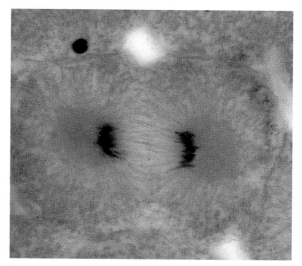

Telophase

The chromosomes lengthen and become less distinct. The cell membrane forms between the forming daughter cells.

(All photographs on this page are 800X.)

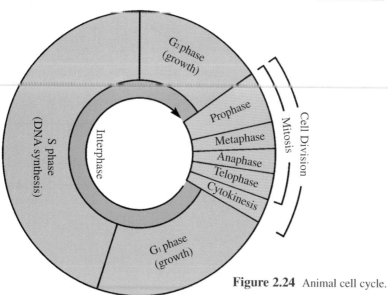

Figure 2.24 Animal cell cycle.

Histology

3

While it is true that cells comprise the basic structural and functional units of the body, the cells in a multicellular organism, such as a human, are so specialized that they do not function independently. *Tissues* are aggregations of similar cells that perform specific functions. *Histology* is the science concerned with the study of cells. Both *cytology*, the study of cells, and histology are actually microscopic anatomy. Although cytologists and histologists utilize many different techniques to study cells and tissues, basically only two kinds of microscopes are used the view the prepared specimens. *Light microscopy* is used for the general observation of cellular and tissue structure, and *electron microscopy* permits observation of the fine details of the specimens.

In electron microscopy, a beam of electrons is passed through an object in a procedure called *transmission electron microscopy (TEM)*, or the beam is reflected off the surface of an object in a procedure called *scanning electron microscopy (SEM)*. In both cases, the electron beam is magnified with electromagnets. The depth of focus of SEM is much greater than it is with SEM, producing a clear three-dimensional image of cellular or tissue structure. The magnification ability of SEM, however, is not as great as that of TEM.

The tissues of the body are classified into four principal types, determined by structure and function: 1) *epithelial tissue* covers body and organ surfaces, lines body and luminal (hollow portion of body tubes) cavities, and forms various glands; 2) *connective tissue* binds, supports, and protects body parts; 3) *muscle tissue* contracts to produce movements; and 4) *nervous tissue* initiates and transmits nerve impulses from one body part to another.

Epithelial tissue is classified by the number of layers of cells and the shape of the cells along the exposed surface. A *simple epithelial tissue* is made up of a single layer of cells. A *stratified epithelial tissue* is made up of layers of cells. The basic shapes of the exposed cells are: *squamous*, or flattened; *cuboidal*, or cube-shaped; and *columnar*, or elongated.

Connective tissue is classified according to the characteristics of the *matrix*, or binding material between the similar cells. The classification of connective tissue is not exact, but the following is a commonly accepted scheme of classification:

A. Embryonic connective tissue

B. Connective tissue proper
   1. Loose (areolar) connective tissue
   2. Dense regular connective tissue
   3. Dense irregular connective tissue
   4. Elastic connective tissue
   5. Reticular connective tissue
   6. Adipose tissue

C. Cartilage
   1. Hyaline cartilage
   2. Fibrocartilage cartilage
   3. Elastic cartilage

D. Bone tissue

E. Blood (vascular tissue)

Muscle tissue is responsible for the movement of materials through the body, the movement of one part of the body with respect to another, and for locomotion. The three kinds of muscle tissue are *smooth, cardiac,* and *skeletal.* The fibers in all three kinds are adapted to contract in response to stimuli.

Nervous tissue is composed of *neurons*, which respond to stimuli and conduct impulses to and from all body organs, and *neuroglia*, which functionally support and physically bind neurons.

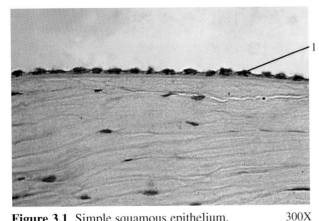

Figure 3.1 Simple squamous epithelium.   300X
1. Single layer of flattened cells

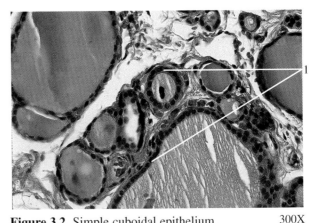

Figure 3.2 Simple cuboidal epithelium.   300X
1. Single layer of cells with round nuclei

19

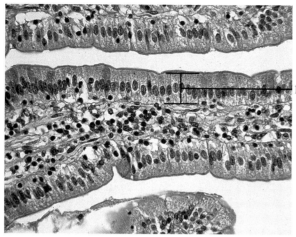

Figure 3.3 Simple columnar epithelium.   300X
1. Single layer of cells with oval nuclei

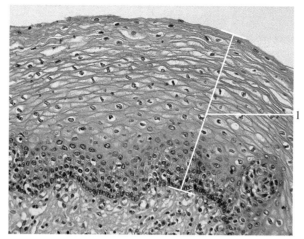

Figure 3.4 Stratified squamous epithelium.   200X
1. Multiple layers of cells, which are flattened at the
upper layer

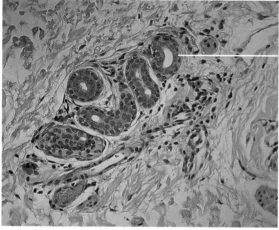

Figure 3.5 Stratified cuboidal epithelium.   200X
1. Two layers of cells with round nuclei

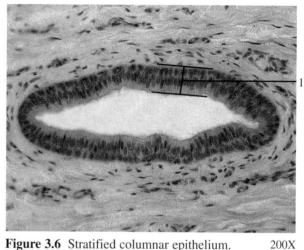

Figure 3.6 Stratified columnar epithelium.   200X
1. Two layers of cells with oval nuclei

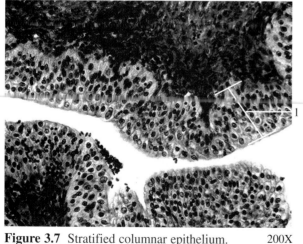

Figure 3.7 Stratified columnar epithelium.   200X
1. Cells are balloon-like at surface

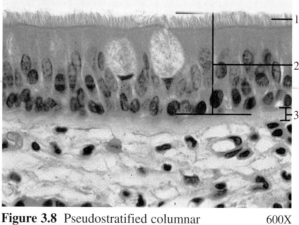

Figure 3.8 Pseudostratified columnar   600X
epithelium.
1. Cilia
1. Pseudostratified columnar epithelium
2. Basement membrane

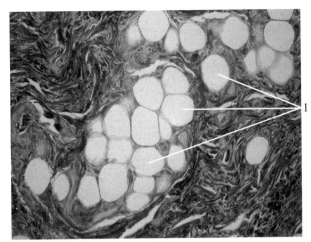

Figure 3.9 Adipose connective tissue.   200X
1. Adipocytes (adipose cells)

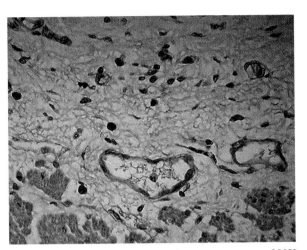

Figure 3.10 Loose connective tissue.   200X

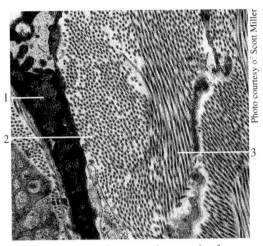

Photo courtesy of Scott Miller

Figure 3.11 An electron micrograph of
collagenous fibers in loose connective tissue.
1. Fibroblast
2. Cross section of bundle of collagenous fibers
3. Longitudinal section of bundle of collagenous fibers

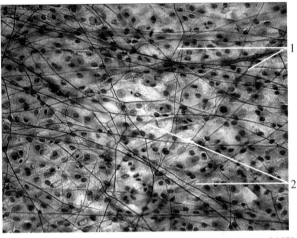

Figure 3.12 Loose connective tissue stained for   200X
fibers.
1. Elastic fibers (black)
2. Collagen fibers (pink)

Figure 3.13 Dense regular connective tissue.   200X
1. Nuclei of fibroblasts arranged in parallel rows

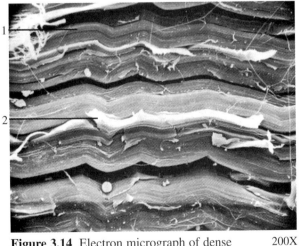

Figure 3.14 Electron micrograph of dense   200X
regular connective tissue.
1. Collagenous fiber
2. Fibroblast

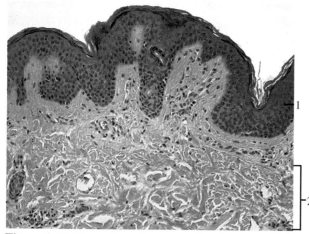

Figure 3.15 Dense irregular connective tissue.   200X
1. Epidermis
2. Dense irregular connective tissue (reticular layer of dermis)

Figure 3.16 Electron micrograph of dense   200X
irregular connective tissue.
1. Collagenous fibers

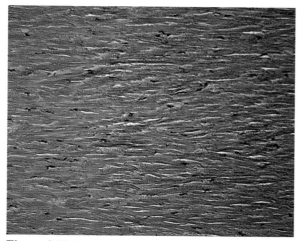

Figure 3.17 Dense irregular connective tissue.   300X

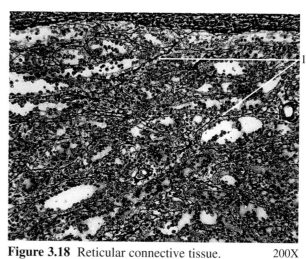

Figure 3.18 Reticular connective tissue.   200X
1. Reticular fibers

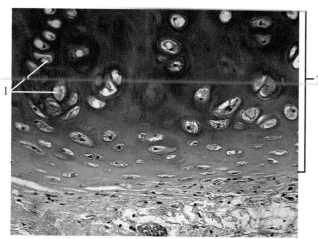

Figure 3.19 Hyaline cartilage.   200X
1. Chondrocytes
2. Hyaline cartilage

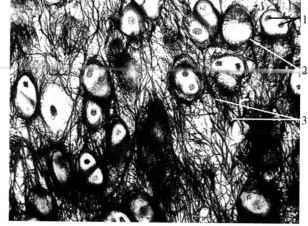

Figure 3.20 Elastic cartilage.   150X
1. Chondrocytes   3. Elastic fibers
2. Lacunae

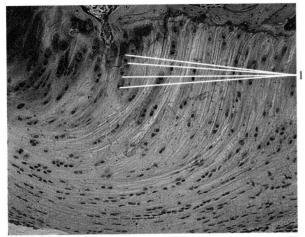

Figure 3.21 Fibrocartilage.   150X
1. Chondrocytes arranged in a row

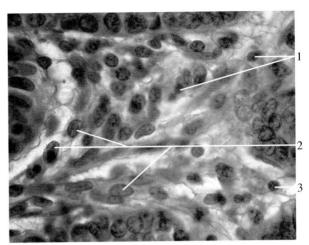

Figure 3.22 Cells of connective tissue.   600X
1. Eosinophils   3. Lymphocyte
2. Fibroblasts

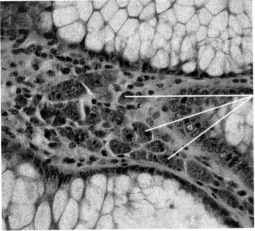

Figure 3.23 Cells of connective tissue.   300X
1. Macrophages

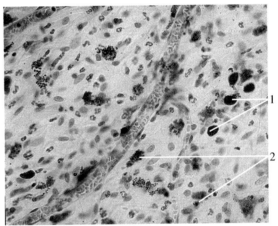

Figure 3.24 Cells of connective tissue,   150X
special preparation.
1. Mast cells   2. Macrophages

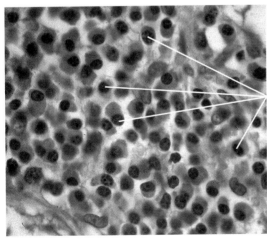

Figure 3.25 Cells of connective tissue.   600X
1. Plasma cells

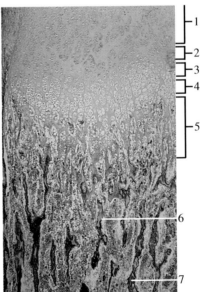

Figure 3.26 Endochondral   40X
bone formation.

1. Zone of reserve
   cartilage
2. Zone of proliferation
3. Zone of hypertrophy
4. Zone of calcification
5. Zone of resorption
6. Calcified cartilage
   (blue)
7. Spicule of bone (red)

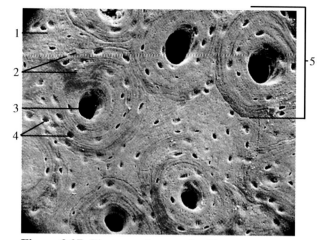

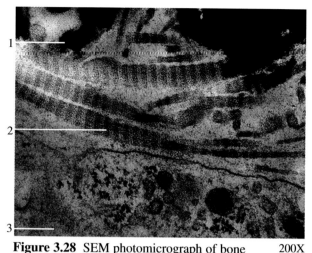

Figure 3.27 Electron micrograph of bone   200X
tissue.
1. Interstitial lamellae   4. Lacunae
2. Lamellae   5. Osteon (haversian system)
3. Central canal
   (haversian canal)

Figure 3.28 SEM photomicrograph of bone   200X
tissue formation.
1. Bone mineral (calcium salts stain black)
2. Collagenous filament (distinct banding pattern)
3. Collagen secreting osteoblasts

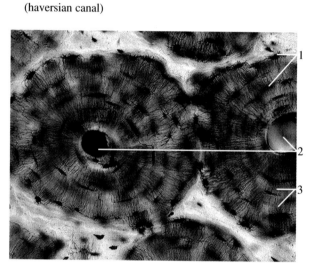

Figure 3.29 Cross section of two osteons.   200X
1. Lacunae   3. Lamellae
2. Central (haversian) canals

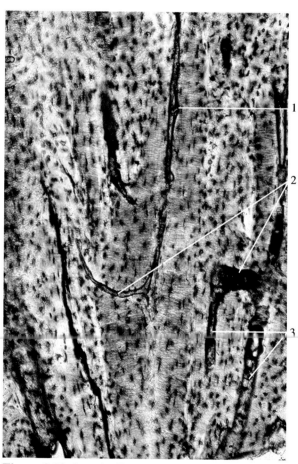

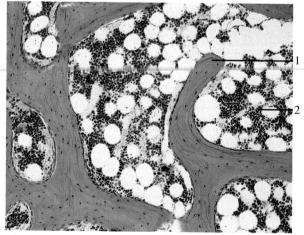

Figure 3.31 Spongy bone.   120X
1. Spicule of bone
2. Marrow (includes adipose cells)

Figure 3.30 Longitudinal section of osteons.   100X
1. Central (haversian) canal
2. Perforating (Volkman's) canals
3. Central (haversian) canals

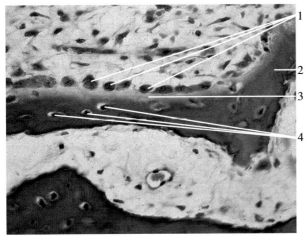

Figure 3.32 Osteoblasts.   375X
1. Osteoblasts   3. Osteoid
2. Bone   4. Osteocytes

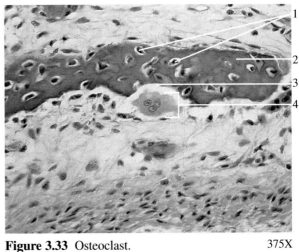

Figure 3.33 Osteoclast.   375X
1. Osteocytes   4. Osteoclast in Howship's
2. Bone   lacuna
3. Howship's lacuna

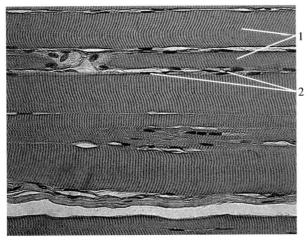

Figure 3.34 Longitudinal section of skeletal   250X
muscle tissue.
1. Skeletal muscle cells, note striations
2. Multiple nuclei in periphery of cell

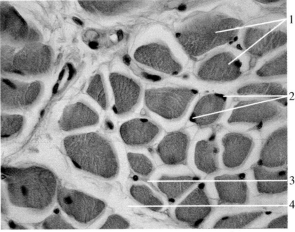

Figure 3.35 Cross section of skeletal   400X
muscle tissue.
1. Skeletal muscle cells
2. Nuclei in periphery of cell
3. Endomysium (surrounds cells)
4. Perimysium (surrounds bundles of cells)

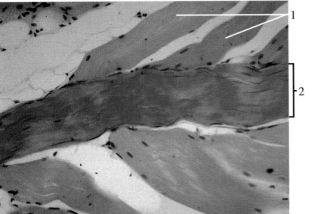

Figure 3.36 Attachment of skeletal muscle to   200X
tendon.
1. Skeletal muscle
2. Dense regular connective tissue (tendon)

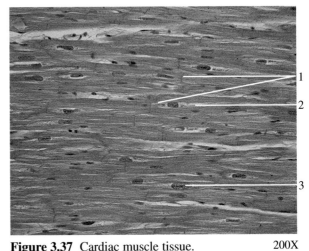

Figure 3.37 Cardiac muscle tissue.   200X
1. Intercalated discs
2. Light-staining perinuclear sarcoplasm
3. Nucleus in center of cell

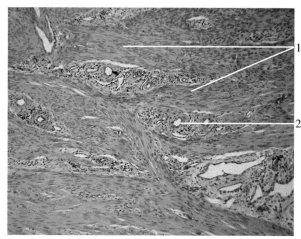

Figure 3.38 Smooth muscle tissue.   75X
1. Smooth muscle
2. Blood vessel

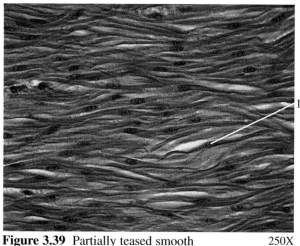

Figure 3.39 Partially teased smooth   250X
muscle tissue.
1. Nucleus of individual cell

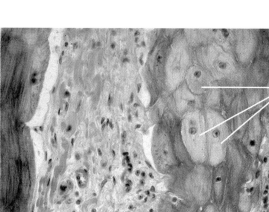

Figure 3.40 Purkinje fibers.   400X
1. Purkinje fibers

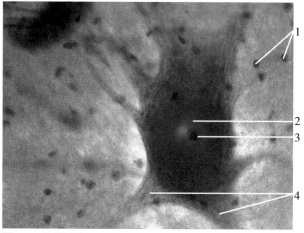

Figure 3.41 Neuron smear.   400X
1. Nuclei of surrounding   3. Nucleolus of neuron
   neuroglial cells   4. Dendrites of neuron
2. Nucleus of neuron

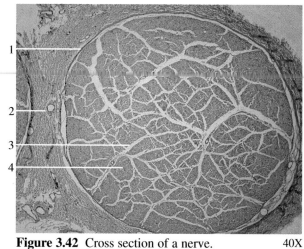

Figure 3.42 Cross section of a nerve.   40X
1. Perineurium   3. Endoneurium
2. Epineurium   4. Bundle of axons

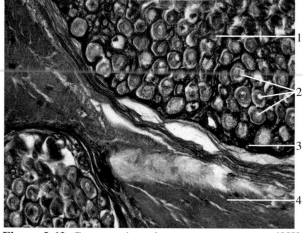

Figure 3.43 Cross section of a nerve.   400X
1. Endoneurium   3. Perineurium
2. Axons   4. Epineurium

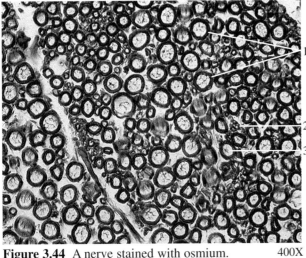

Figure 3.44 A nerve stained with osmium. 400X
1. Myelin sheath
2. Endoneurium
3. Axon

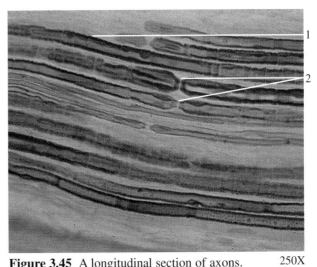

Figure 3.45 A longitudinal section of axons. 250X
1. Myelin sheath
2. Neurofibril nodes (nodes of Ranvier)

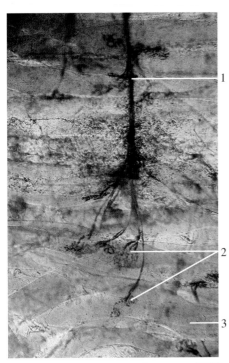

Figure 3.46 Neuromuscular 250X
junction.

1. Motor nerve
2. Motor end plates
3. Skeletal muscle fiber

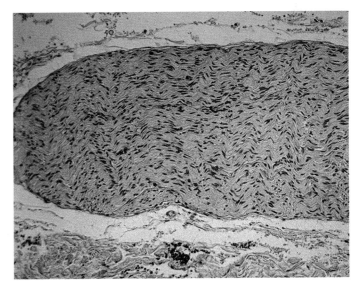

Figure 3.47 Cross section of unmyelinated nerve. 100X

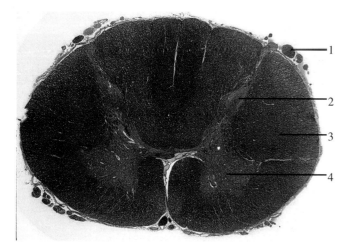

Figure 3.48 Cross section of the spinal cord.   7X
1. Posterior (dorsal) root of spinal nerve
2. Posterior (dorsal) horn (gray matter)
3. Spinal cord tract (white matter)
4. Anterior (ventral) horn (gray matter)

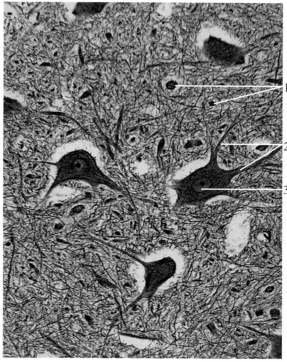

Figure 3.49 Motor neurons from spinal cord.   250X
1. Neuroglia cells
2. Dendrites
3. Nucleus

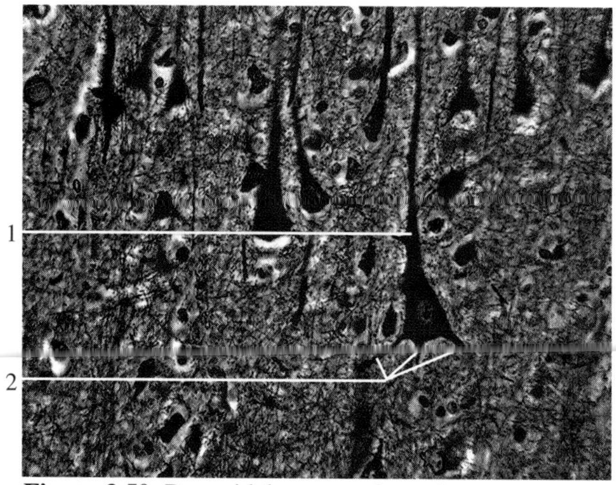

Figure 3.50 Pyramidal neurons from the   250X
cerebral cortex.
1. Apical dendrite
2. Other dendrites

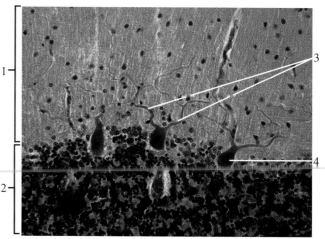

Figure 3.51 Purkinje neurons from the   200X
cerebellum.
1. Molecular layer of cerebellar cortex
2. Granular layer of cerebellar cortex
3. Dendrites of Purkinje cell
4. Purkinje cell body

Integumentary System

The integumentary system consists of the *integument*, or *skin*, and its associated *hair*, *glands*, and *nails* (fig. 4.1). The skin is composed of an outer *epidermis* consisting of four or five layers and a *dermis* consisting of two layers. The *hypodermis* (*subcutaneous tissue*) connects the skin to the underlying organs.

The stratified squamous epithelium of the epidermis is divisible into five *strata*, or layers. From superficial to deep, they are the *stratum corneum*, the *stratum lucidum* (only in skin of the palms and soles), the *stratum granulosum*, the *stratum spinosum*, and the *stratum basale*. The strata basale and spinosum undergo mitosis (cell division) and are collectively called the *stratum germinativum*. Pigments are found in the stratum germinativum and the protein *keratin* is found in all but the deepest epidermal layers. Both are protective. The stratum corneum is *cornified* (hardened and scale-like) for further protection.

The dermis is divisible into the *stratum papillarosum* (papillary layer) and the *stratum reticularosum* (reticular layer). The hypodermis is the deep, binding layer of connective tissue.

The skin provides several important functions, including: 1) protection of the body from disease and external injury. Keratin and an acidic oily secretion on the surface protect the skin from water and microorganisms. Cornification protects against abrasion, and *melanin* (a dark pigment) is a barrier to UV light; 2) regulation of body fluids and temperatures by radiation, convection, and the antagonistic effects of sweating and shivering; 3) permits the absorption of some UV light, respiratory gases, steroids, and fat-soluble vitamins; 4) synthesizes melanin and keratin, which remain in the skin, and vitamin D, which is used elsewhere in the body; 5) sensory reception provided through cutaneous receptors throughout the dermis and hypodermis; and 6) development and growth of hair and certain exocrine glands.

Formed prenatally as invaginations of the epidermis into the dermis, hair, glands and nails provide protection to the skin. Each hair develops in a *hair follicle* and is protective against sunlight and mild abrasions. Integumentary glands are classified as *sebaceous* (oil secreting), *sudoriferous* (sweat), and *ceruminous* (wax-producing). (*Mammary glands* are specialized sweat glands that produce milk in a lactating female.) A nail protects the terminal end of each digit. The fingernails also aid in picking up objects and scratching.

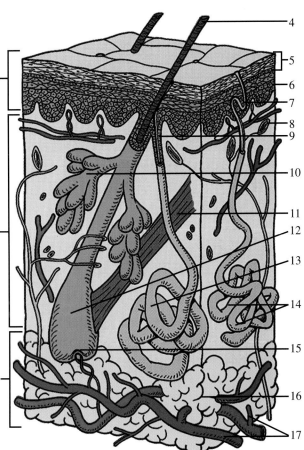

Figure 4.1 The skin and certain epidermal structures.

1. Epidermis	10. Sebaceous gland
2. Dermis	11. Arrector pili muscle
3. Hypodermis	12. Hair follicle
4. Shaft of hair	13. Apocrine sweat gland
5. Stratum corneum	14. Eccrine sweat gland
6. Stratum basale	15. Bulb of hair
7. Sweat duct	16. Adipose tissue
8. Sensory receptor	17. Cutaneous blood vessels
9. Sweat duct	

Figure 4.2 The gross structure of the skin and underlying fascia.
1. Epidermis   3. Hypodermis   5. Muscle
2. Dermis   4. Fascia

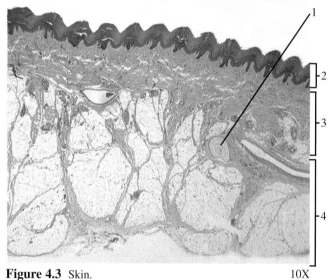

Figure 4.3 Skin.   10X
1. Lamellated (Pacinian) corpuscle   3. Dermis
2. Epidermis   4. Hypodermis

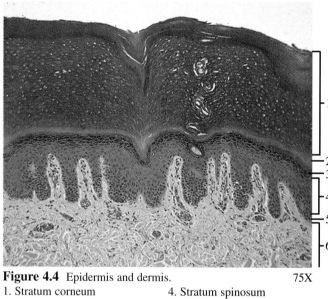

Figure 4.4 Epidermis and dermis.   75X
1. Stratum corneum   4. Stratum spinosum
2. Stratum lucidum   5. Stratum basale
3. Stratum granulosum   6. Dermis

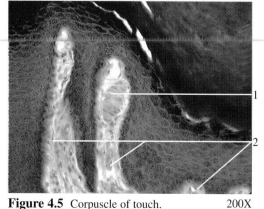

Figure 4.5 Corpuscle of touch.   200X
1. Corpuscle of touch (Meissner's corpuscle)
2. Dermal papillae

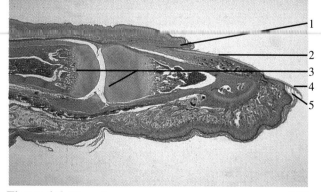

Figure 4.6 Fingertip.   10X
1. Eponychium   4. Free border of nail
2. Nail plate   5. Hyponychium
3. Phalanges

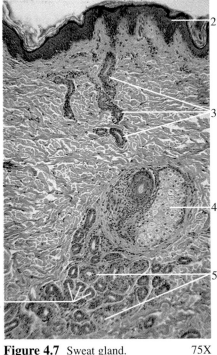

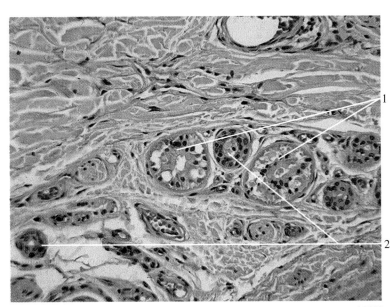

Figure 4.7 Sweat gland.   75X
1. Excretory portion of sweat gland
2. Epidermis
3. Excretory duct of sweat gland
   (coiling toward surface)
4. Sebaceous gland
5. Secretory portion of sweat gland

Figure 4.8 Sweat gland.   200X
1. Secretory portion (large diameter with light–staining columnar cells)
2. Excretory portion (small diameter with dark–staining stratified cuboidal cells)

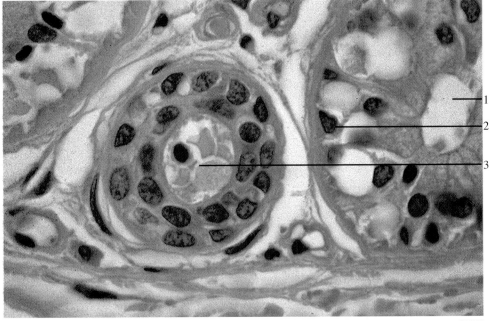

Figure 4.9 Sweat gland.   600X
1. Lumen of secretory portion   3. Lumen of excretory portion
2. Myoepithelial cell

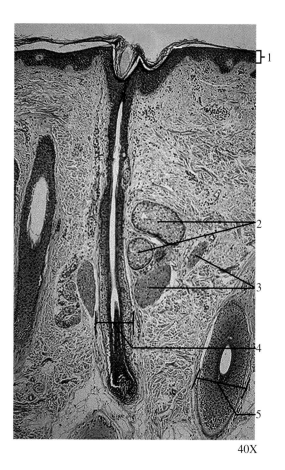

40X

Figure 4.10 Hair follicle.
1. Epidermis
2. Sebaceous glands
3. Arrector pili muscle
4. Hair follicle
5. Hair follicle (oblique cut)

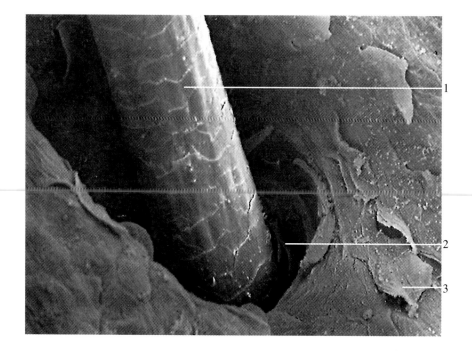

Figure 4.11 An electron micrograph of a hair emerging from a hair follicle.
1. Shaft of hair (note the scale–like pattern)
2. Hair follicle
3. Epithelial cell from stratum corneum

Skeletal System: Axial Portion 5

The skeletal system of an adult human is composed of approximately 206 bones—the number varies from person to person depending on genetic variations. Some adults have extra bones in the skull called *sutural (wormian) bones*. Additional bones may develop in tendons as the tendons move across a joint. Bones formed this way are called *sesamoid bones*, and the patella (kneecap) is an example.

The skeleton is divided into axial and appendicular portions (table 5.1). The *axial skeleton* consists of the bones that form the axis of the body and that support and protect the organs of the head, neck, and trunk. The axial skeleton includes the bones of the skull, auditory ossicles, hyoid bone, vertebral column, and rib cage.

The *appendicular skeleton* (see chapter 6) is composed of the bones of the upper and lower extremities and the bony girdles, which anchor the appendages to the axial skeleton. The appendicular skeleton includes the bones of the pectoral girdle, upper extremities, pelvic girdle, and lower extremities.

The mechanical functions of the bones of the skeleton include the support and protection of softer body tissues and organs. Also, certain bones function as levers during body movement. The metabolic functions of bones include *hemopoiesis*, or manufacture of blood cells, and mineral storage. Calcium and phosphorus are the two principal minerals stored within bone, and give bone its rigidity and strength.

The bones of the skeleton are classified into four principal types on the basis of shape rather than size. The four classes of bones are *long bones*, *short bones*, *flat bones* and *irregular bones* (fig. 5.1).

Table 5.1 Classifications of the Bones of the Adult skeleton

Appendicular Skeleton		Axial Skeleton	
Pectoral girdle— 4 bones	**Pelvic girdle— 2 bones**	**Auditory ossicles— 6 bones**	**skull - 22 bones**
scapula (2)	os coxae (2) (each contains 3	malleus (2)	*14 facial bones*
clavicle (2)	fused bones: ilium, ischium,	incus (2)	maxilla (2)
Upper extremities— 60 bones	and pubis)	stapes (2)	palatine bone (2)
	Lower extremities— 60 bones	**Hyoid—1 bone**	zygomatic bone (2)
		Vertebral column —26 bones	lacrimal bone (2)
metacarpal bone (10)	femur (2)		nasal bone (2)
humerus (2)	tarsal bone (14)	cervical vertebra (7)	vomer (1)
carpal bone (16)	tibia (2)	thoracic vertebra (12)	inferior nasal concha (2)
radius (2)	metatarsal bone (10)	lumbar vertebra (5)	mandible (1)
metacarpal bone (10)	fibula (2)	sacrum (1) (5 fused bones)	*8 cranial bones*
ulna (2)	phalanx (28)	coccyx (1) (3 to 5 fused bones)	frontal bone (1)
phalanx (28)	patella (2)	**Rib cage—25 bones**	parietal bone (2)
		rib (24)	occipital bone (1)
		sternum (1)	temporal bone (2)
			sphenoid bone (1)
			ethmoid bone (1)

Short bone

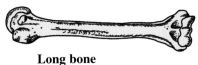

Long bone

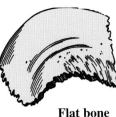

Flat bone

Irregular bone

Figure 5.1 Shapes of bones.

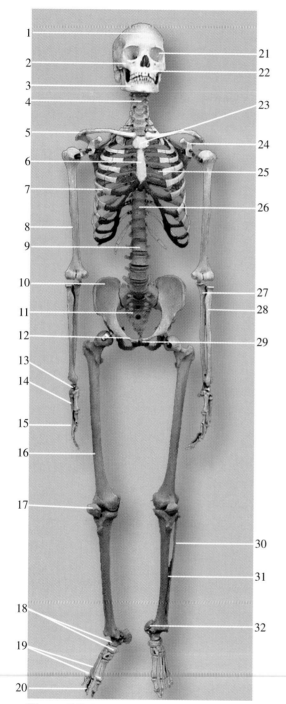

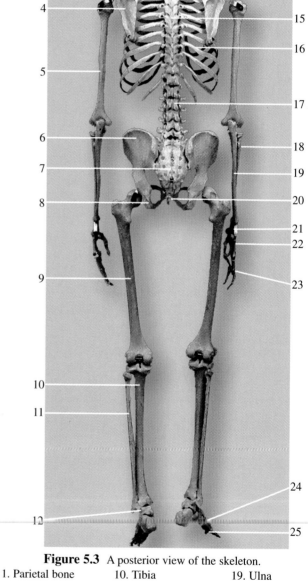

Figure 5.2 An anterior view of the skeleton.

1. Frontal bone	12. Pubis	23. Manubrium
2. Zygomatic bone	13. Carpal bones	24. Scapula
3. Mandible	14. Metacarpal bones	25. Costal cartilage
4. Cervical vertebra	15. Phalanges	26. Thoracic vertebra
5. Clavicle	16. Femur	27. Ulna
6. Body of sternum	17. Patella	28. Radius
7. Rib	18. Tarsal bones	29. Symphysis pubis
8. Humerus	19. Metatarsal bones	30. Fibula
9. Lumbar vertebra	20. Phalanges	31. Tibia
10. Ilium	21. Orbit	32. Calcaneus
11. Sacrum	22. Maxilla	

Figure 5.3 A posterior view of the skeleton.

1. Parietal bone	10. Tibia	19. Ulna
2. Occipital bone	11. Fibula	20. Coccyx
3. Cervical vertebra	12. Tarsal bones	21. Carpal bones
4. Scapula	13. Mandible	22. Metacarpal bones
5. Humerus	14. Clavicle	23. Phalanges
6. Ilium	15. Thoracic vertebra	24. Metatarsal bones
7. Sacrum	16. Rib	25. Phalanges
8. Ischium	17. Lumbar vertebra	
9. Femur	18. Radius	

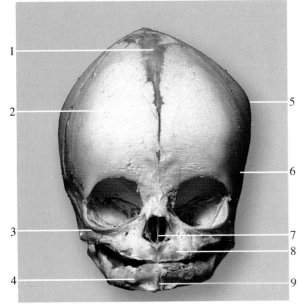

Figure 5.4 An anterior view of the fetal skull.

1. Anterior fontanel
2. Frontal bone
3. Zygomatic bone
4. Mandible
5. Parietal bone
6. Anterolateral fontanel
7. Nasal septum
8. Maxilla
9. Mental symphysis

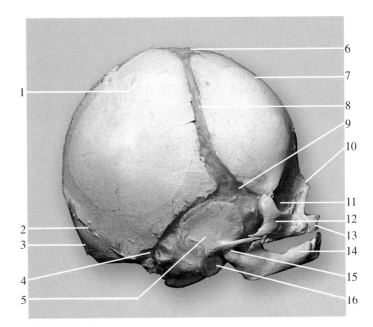

Figure 5.5 A lateral view of the fetal skull.

1. Parietal bone
2. Area of lambdoidal suture
3. Occipital bone
4. Posterolateral fontanel
5. Temporal bone
6. Anterior fontanel
7. Frontal bone
8. Area of coronal suture
9. Anterolateral fontanel
10. Nasal bone
11. Sphenoid bone
12. Zygomatic bone
13. Maxilla
14. Mandible
15. Condylar process
16. External acoustic meatus

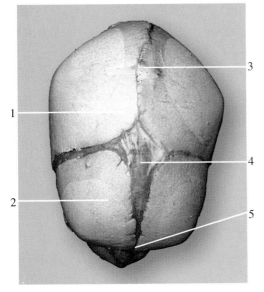

Figure 5.6 A superior view of the fetal skull.

1. Parietal bone
2. Frontal bone
3. Area of sagittal suture
4. Anterior fontanel
5. Synostosis

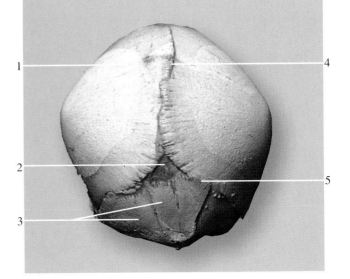

Figure 5.7 A posterior view of the fetal skull.

1. Parietal bone
2. Posterior fontanel
3. Occipital bone
4. Area of sagittal suture
5. Area of lambdoidal suture

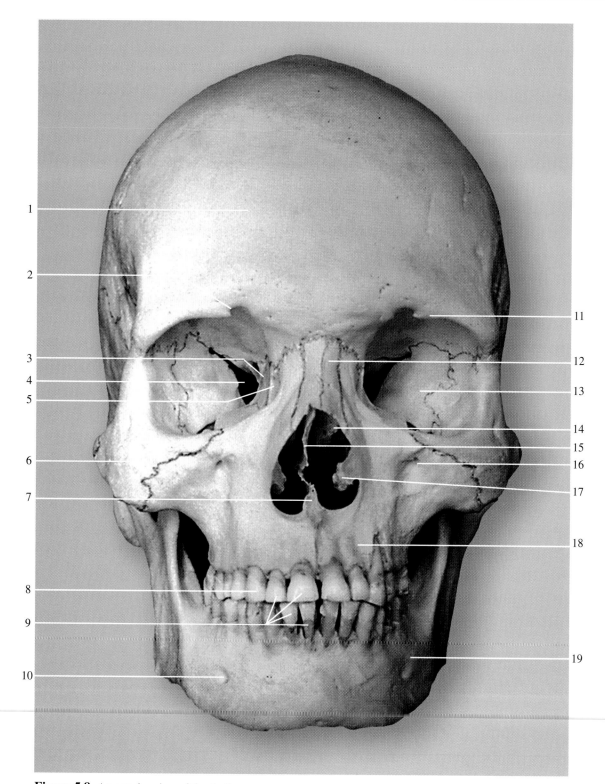

Figure 5.8 An anterior view of the skull.

1. Frontal bone
2. Supraorbital foramen
3. Orbital plate of ethmoid bone
4. Superior orbital fissure
5. Lacrimal bone
6. Zygomatic bone
7. Vomer
8. Canine
9. Incisors
10. Mental foramen
11. Supraorbital margin
12. Nasal bone
13. Sphenoid bone
14. Middle nasal concha of ethmoid bone
15. Perpendicular plate of ethmoid bone
16. Infraorbital foramen
17. Inferior nasal concha
18. Maxilla
19. Mandible

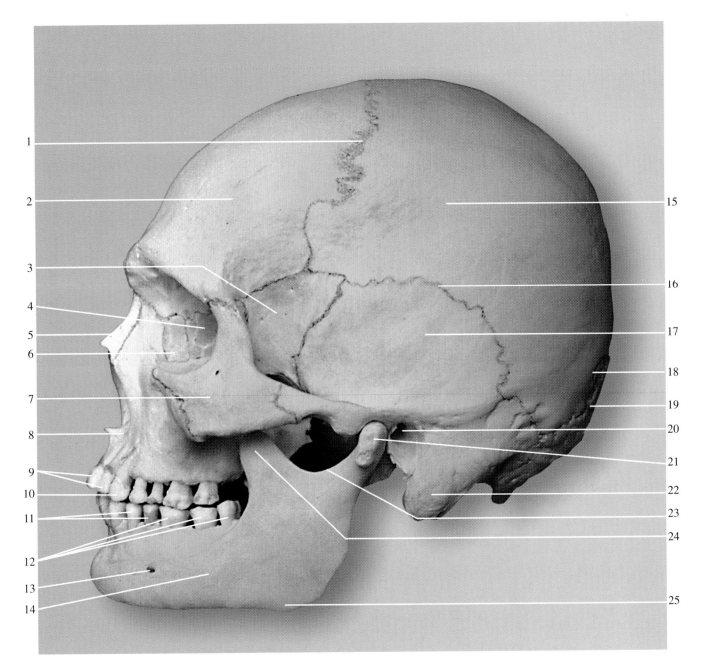

Figure 5.9 A lateral view of the skull.

1. Coronal suture
2. Frontal bone
3. Sphenoid bone
4. Orbital plate of ethmoid bone
5. Nasal bone
6. Lacrimal bone
7. Zygomatic bone
8. Maxilla
9. Incisors
10. Canine
11. Premolars
12. Molars
13. Mental Foramen
14. Mandible
15. Parietal bone
16. Squamosal suture
17. Temporal bone
18. Lambdoidal suture
19. Occipital bone
20. External acoustic meatus
21. Condylar process of mandible
22. Mastoid process of temporal bone
23. Mandibular notch
24. Coronoid process of mandible
25. Angle of mandible

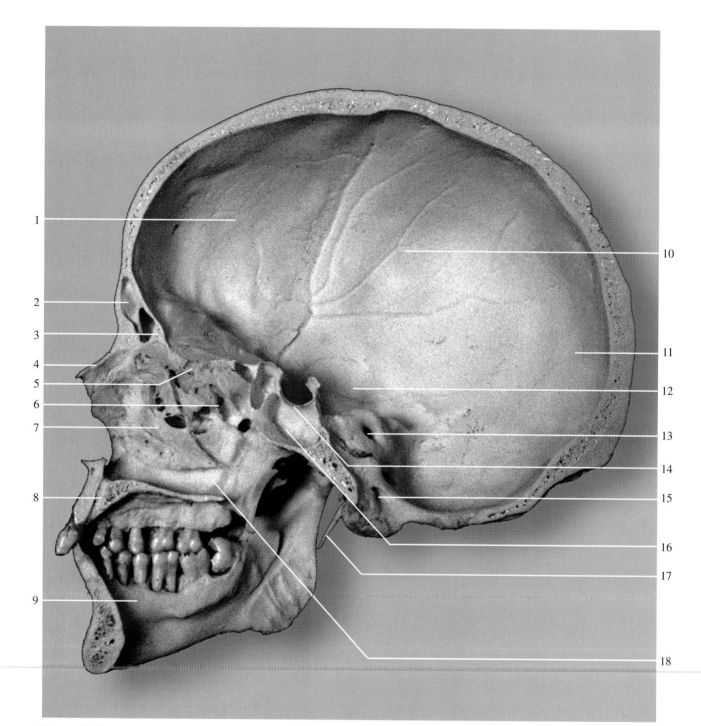

Figure 5.10   A sagittal view of the skull.

1. Frontal bone	7. Nasal concha	13. Internal acoustic meatus
2. Frontal sinus	8. Maxilla	14. Sella turcica
3. Crista galli of ethmoid bone	9. Mandible	15. Hypoglossal canal
4. Nasal bone	10. Parietal bone	16. Sphenoidal sinus
5. Cribriform plate of ethmoid bone	11. Occipital bone	17. Styloid process of temporal bone
6. Ethmoidal sinus	12. Temporal bone	18. Vomer

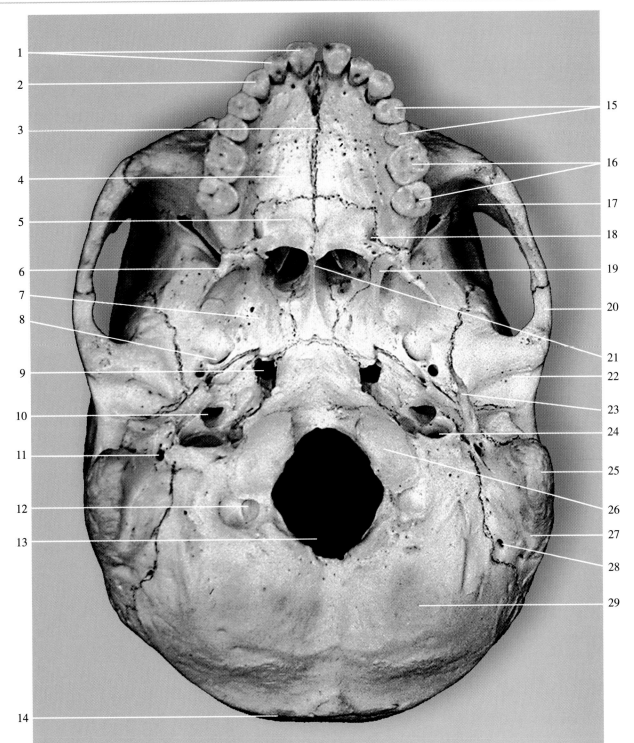

Figure 5.11 An inferior view of the skull.

1. Incisors
2. Canine
3. Median palatine suture
4. Maxilla
5. Palatine bone
6. Lateral pterygoid process of sphenoid bone
7. Sphenoid bone
8. Foramen ovale
9. Foramen lacerum
10. Carotid canal
11. Stylomastoid foramen
12. Condyloid canal
13. Foramen magnum
14. Superior nuchal line
15. Premolars
16. Molars
17. Zygomatic bone
18. Greater palatine foramen
19. Medial pterygoid process of sphenoid bone
20. Zygomatic arch
21. Vomer
22. Mandibular fossa
23. Styloid process of temporal bone
24. Jugular fossa
25. Mastoid process of temporal bone
26. Occipital condyle
27. Temporal bone
28. Mastoid foramen
29. Occipital bone

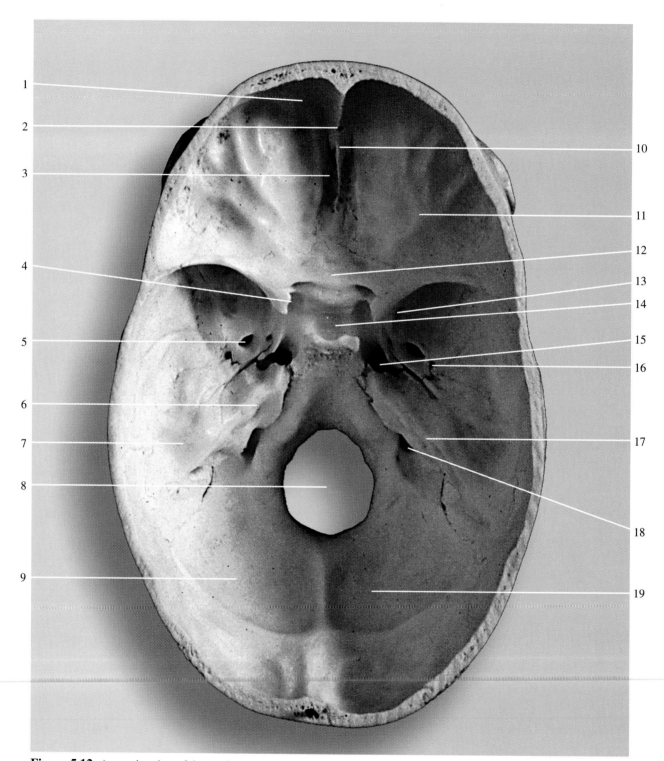

Figure 5.12 A superior view of the cranium.

1. Frontal bone
2. Foramen cecum
3. Cribriform plate of ethmoid bone
4. Optic canal
5. Foramen ovale
6. Petrous part of temporal bone
7. Temporal bone

8. Foramen magnum
9. Occipital bone
10. Crista galli of ethmoid bone
11. Anterior cranial fossa
12. Sphenoid bone
13. Foramen rotundum
14. Sella turcica of sphenoid bone

15. Foramen lacerum
16. Foramen spinosum
17. Internal acoustic meatus
18. Jugular foramen
19. Posterior cranial fossa

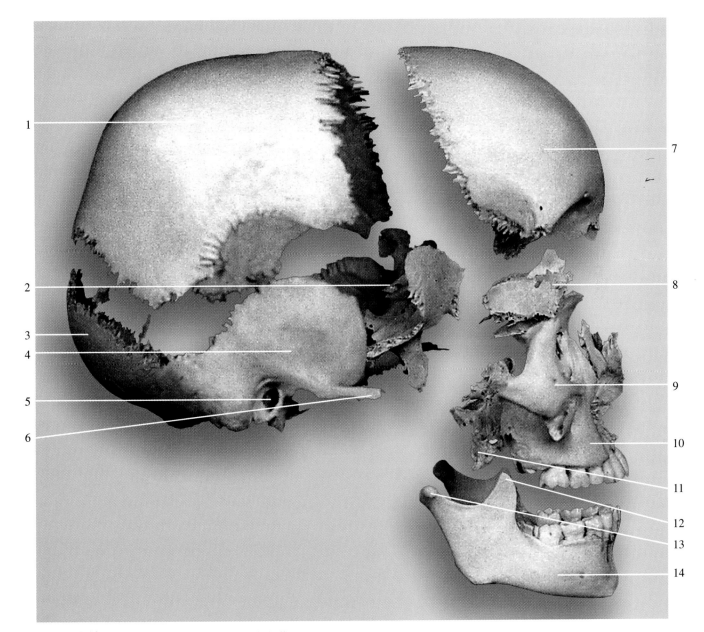

Figure 5.13 A lateral view of a disarticulated skull.

1. Parietal bone
2. Sphenoid bone
3. Occipital bone
4. Temporal bone
5. External acoustic meatus
6. Zygomatic process of temporal bone
7. Frontal bone
8. Ethmoid bone
9. Zygomatic bone
10. Maxilla
11. Palatine bone
12. Coronoid process of mandible
13. Condylar process of mandible
14. Mandible

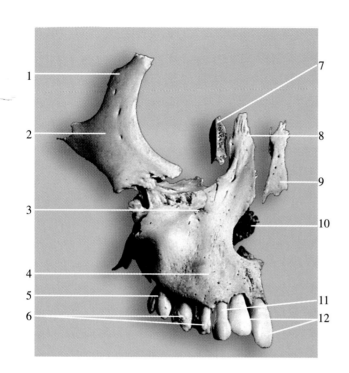

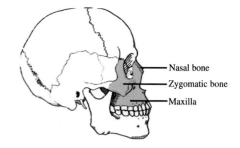

Figure 5.14 Bones of the left facial region.
1. Orbital process of zygomatic bone
2. Zygomatic bone
3. Infraorbital foramen
4. Maxilla
5. Molar
6. Premolars
7. Lacrimal bone
8. Frontal process of maxilla
9. Nasal bone
10. Inferior nasal concha
11. Canine
12. Incisors

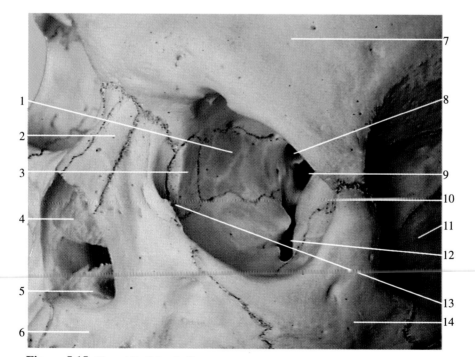

Figure 5.15 The orbit of the skull.
1. Ethmoid bone
2. Nasal bone
3. Lacrimal bone
4. Inferior nasal concha
5. Vomer
6. Maxilla
7. Frontal bone
8. Optic foramen
9. Superior orbital fissure
10. Sphenoid bone
11. Temporal bone
12. Inferior orbital fissure
13. Lacrimal foramen
14. Zygomatic bone

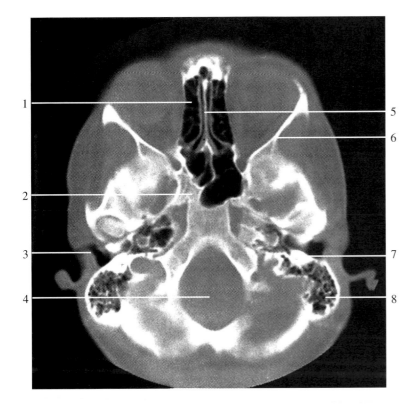

Figure 5.16 A CT transaxial section through the skull.
1. Nasal cavity
2. Sphenoid bone
3. External acoustic meatus
4. Foramen magnum
5. Nasal septum
6. Wall of bony orbit
7. Middle-ear chamber
8. Mastoidal sinus

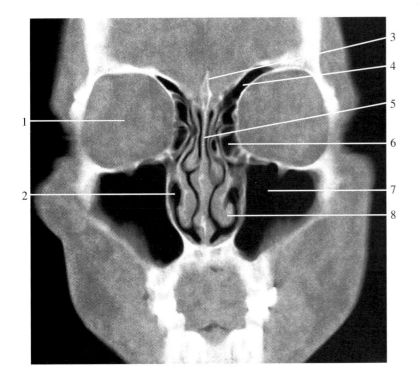

Figure 5.17 A CT image of the nasal cavity and paranasal sinuses.
1. Eyeball in orbit
2. Nasal cavity
3. Crista galli
4. Frontal sinus
5. Perpendicular plate of ethmoid bone
6. Ethmoidal sinus
7. Maxillary sinus
8. Inferior nasal concha

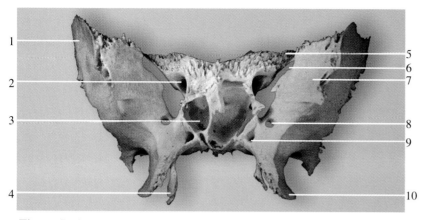

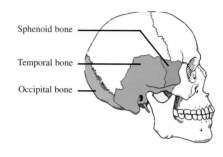

Figure 5.18 An anterior view of the sphenoid bone.

1. Greater wing of sphenoid bone
2. Optic foramen
3. Opening into sphenoidal sinus
4. Medial pterygoid plate
5. Lesser wing of sphenoid bone
6. Superior orbital fissure
7. Orbital surface of greater wing of sphenoid bone
8. Foramen rotundum
9. Pterygoid canal
10. Lateral pterygoid plate

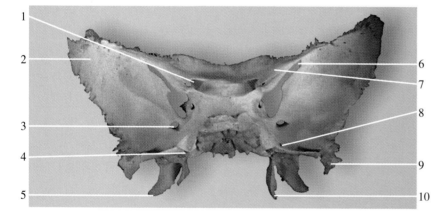

Figure 5.19 A posterior view of the sphenoid bone.

1. Optic foramen
2. Greater wing of sphenoid bone
3. Foramen rotundum
4. Pterygoid canal
5. Lateral pterygoid plate
6. Superior orbital fissure
7. Lesser wing of sphenoid bone
8. Foramen ovale
9. Spine of sphenoid bone
10. Medial pterygoid plate

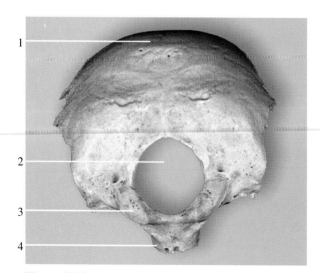

Figure 5.20 An inferior view of the occipital bone.
1. External occipital protuberance
2. Foramen magnum
3. Occipital condyle
4. Pharyngeal tubercle

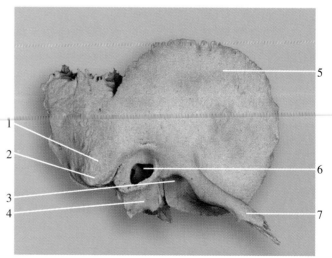

Figure 5.21 A lateral view of the temporal bone.
1. Mastoid part of temporal bone
2. Mastoid process
3. Mandibular fossa
4. Tympanic part of temporal bone
5. Squamous part of temporal bone
6. External acoustic meatus
7. Zygomatic process of temporal bone

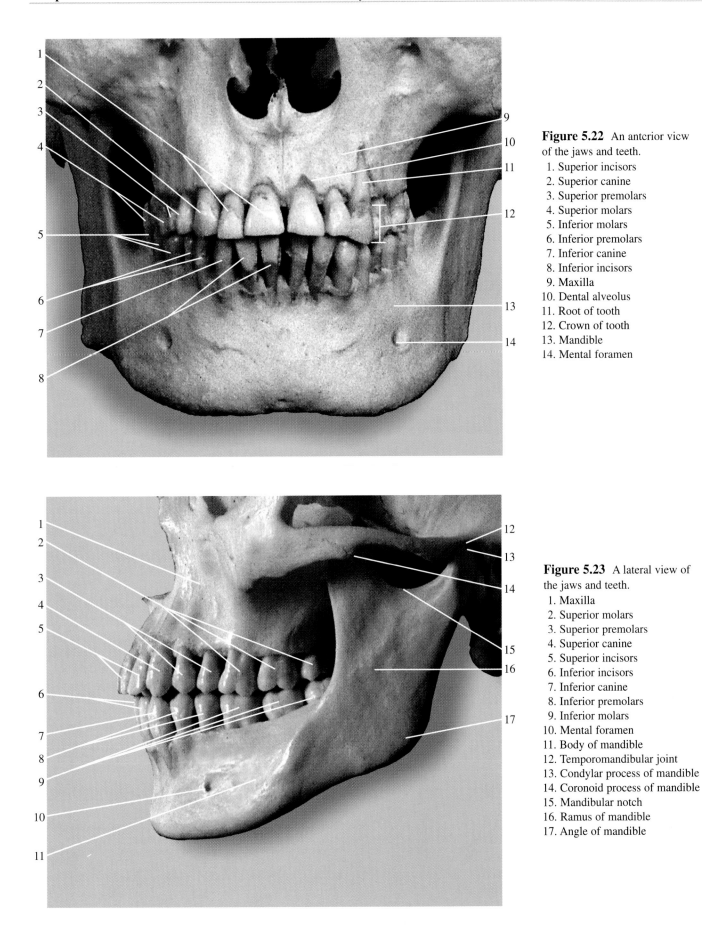

Figure 5.22 An anterior view of the jaws and teeth.
1. Superior incisors
2. Superior canine
3. Superior premolars
4. Superior molars
5. Inferior molars
6. Inferior premolars
7. Inferior canine
8. Inferior incisors
9. Maxilla
10. Dental alveolus
11. Root of tooth
12. Crown of tooth
13. Mandible
14. Mental foramen

Figure 5.23 A lateral view of the jaws and teeth.
1. Maxilla
2. Superior molars
3. Superior premolars
4. Superior canine
5. Superior incisors
6. Inferior incisors
7. Inferior canine
8. Inferior premolars
9. Inferior molars
10. Mental foramen
11. Body of mandible
12. Temporomandibular joint
13. Condylar process of mandible
14. Coronoid process of mandible
15. Mandibular notch
16. Ramus of mandible
17. Angle of mandible

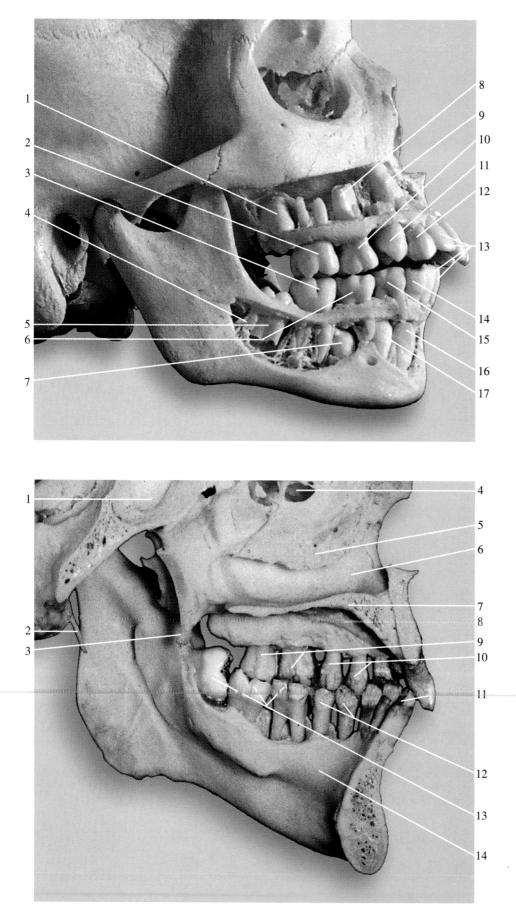

Figure 5.24 Eruption of teeth seen in a dissected skull of a youth (9 to 12 years old).
1. Permanent second molar
2. Permanent first molar
3. Permanent first molar
4. Permanent third molar
5. Permanent second molar
6. Deciduous second premolar
7. Permanent second premolar
8. Permanent second premolar
9. Permanent canine
10. Deciduous second premolar
11. Permanent first premolar
12. Deciduous canine
13. Incisors
14. Deciduous canine
15. Deciduous first premolar
16. Permanent canine
17. Permanent first premolar

Figure 5.25 A medial view of the jaws and teeth.
1. Sphenoidal sinus
2. Styloid process of temporal bone
3. Medial pterygoid process
4. Ethmoidal sinus
5. Inferior nasal concha
6. Inferior meatus
7. Hard palate
8. Maxilla
9. Superior molars
10. Superior premolars
11. Incisors
12. Inferior premolars
13. Inferior molars
   (note impacted wisdom tooth)
14. Mandible

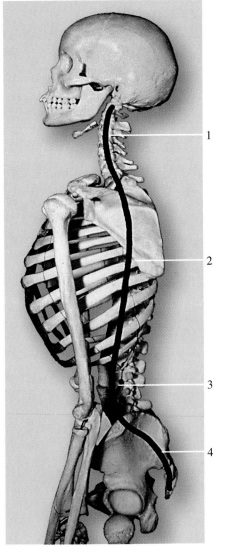

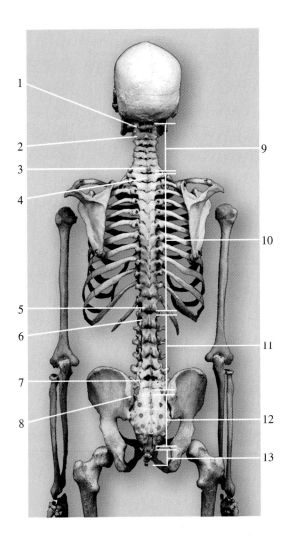

Figure 5.26 The curvature of the vertebral column.
1. Cervical curvature
2. Thoracic curvature
3. Lumbar curvature
4. Pelvic (sacral) curvature

Figure 5.27 A posterior view of the vertebral column.
1. Atlas
2. Axis
3. Seventh cervical vertebra
4. First thoracic vertebra
5. Twelfth thoracic vertebra
6. First lumbar vertebra
7. Fifth lumbar vertebra
8. Sacroiliac joint
9. Cervical vertebrae
10. Thoracic vertebrae
11. Lumbar vertebrae
12. Sacrum
13. Coccyx

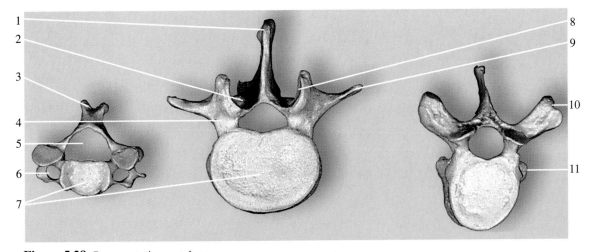

Figure 5.28 Representative vertebrae.
1. Spinous process
2. Lamina
3. Spinous process
   (note that it is bifid)
4. Pedicle
5. Vertebral canal
6. Transverse canal
7. Bodies of vertebrae
8. Superior articular surface
9. Transverse process
10. Costal fovea of transverse process
11. Inferior costal fovea

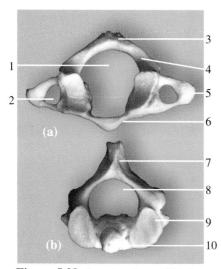

Figure 5.29 Superior views of (a) the atlas and (b) the axis.

1. Vertebral canal
2. Transverse foramen
3. Spinous process of atlas
4. Lamina of neural arch
5. Transverse process of atlas
6. Anterior arch of atlas
7. Spinous process of axis
8. Vertebral canal
9. Superior articular facet for atlas
10. Dens (odontoid process)

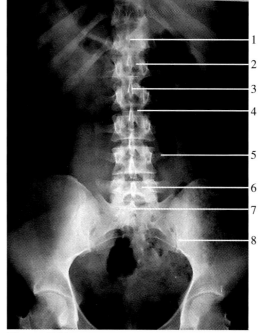

Figure 5.30 A radiograph of the lumbar vertebrae.

1. T12
2. Body of L1
3. Spinous process of L2
4. Intervertebral disc
5. Transverse process of L4
6. Lamina of L5
7. Sacrum
8. Sacroiliac joint

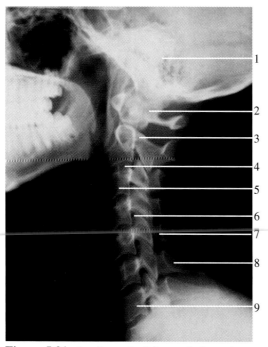

Figure 5.31 A radiograph of the cervical vertebrae.

1. Occipital condyle
2. Atlas
3. Axis
4. Intervetebral disc
5. Body of C3
6. Intervetebral foramen
7. Spinous process of C5
8. Spinous process of C6
9. Body of C7

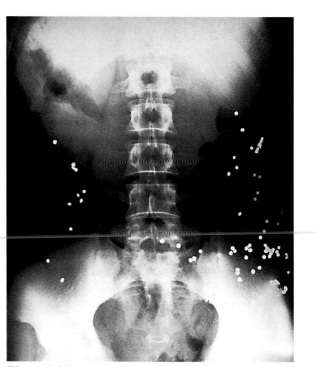

Figure 5.32 A radiograph of the lumbar region showing the locations of pellets from a shotgun wound.

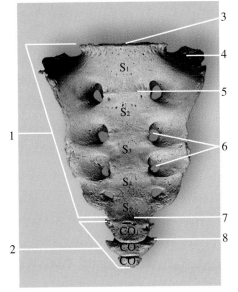

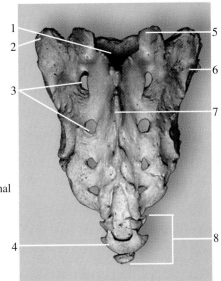

Figure 5.33 An anterior view of the sacrum and coccyx.
1. Sacrum
2. Coccyx
3. Base of sacrum
4. Ala
5. Transverse line
6. Anterior sacral foramina
7. Apex of sacrum
8. Coccygeal cornu

Figure 5.34 A posterior view of the sacrum and coccyx.
1. Superior portion of sacral canal
2. Ala
3. Posterior sacral foramina
4. Coccygeal cornu
5. Superior articular process
6. Auricular surface
7. Median sacral crest
8. Coccyx

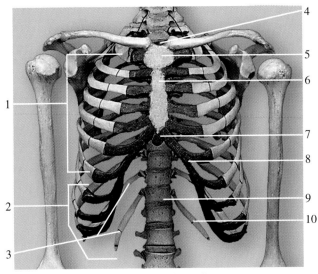

Figure 5.35 An anterior view of the rib cage.
1. True ribs (seven pairs)
2. False ribs (five ribs)
3. Floating ribs (inferior two pairs of false ribs)
4. Jugular notch
5. Manubrium
6. Body of sternum
7. Xiphoid process
8. Costal cartilage
9. Twelfth thoracic vertebra
10. Twelfth rib

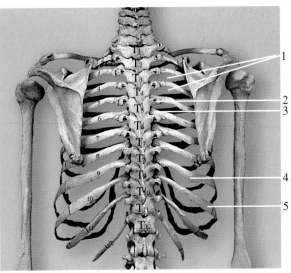

Figure 5.36 A posterior view of the rib cage.
1. Intercostal spaces
2. Transverse process of thoracic vertebra
3. Head of rib
4. Angle of rib
5. Body of rib

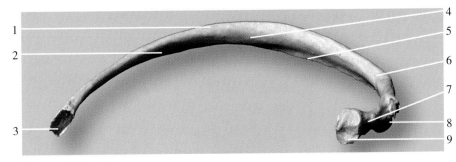

Figure 5.37 A rib.
1. Superior border
2. Body of rib
3. Articulation site for costal cartilage
4. Internal surface
5. Costal groove
6. Angle of rib
7. Neck of rib
8. Articular surface
9. Head of rib

Skeletal System: Appendicular Portion 6

The structure of the *pectoral girdle* and *upper extremities* of the appendicular skeleton is adaptive for freedom of movement and extensive muscle attachment. The structure of the *pelvic girdle* and *lower extremities* is adaptive for support and locomotion.

The pectoral girdle is composed of two *scapulae* and two *clavicles*. The clavicles attach the pectoral girdle to the axial skeleton at the *sternum*. The bones of each upper extremity are the *humerus, ulna, radius*, eight *carpal bones*, five *metacarpal bones*, and fourteen *phalanges*. In addition, two or more *sesamoid bones* at the interphalangeal joints are usually present.

The pelvic girdle, or *pelvis*, is formed by two *ossa coxae* (hip bones) which are united anteriorly by the *symphysis pubis*. The pelvic girdle is attached posteriorly to the sacrum of the vertebral column. Each *os coxae* consists of three separate bones: the *ilium*, the *ischium*, and the *pubis*. These bones are fused in an adult. The bones of each lower extremity are the *femur, patella, tibia, fibula*, seven *tarsal bones*, five *metatarsal bones*, and fourteen *phalanges*. In addition, two or more sesamoid bones at the interphalangeal joints are usually present.

The various surface features used to identify specific bones are presented in table 6.1.

Table 6.1 Surface Features Used to Identify Specific Bones

Articulating surfaces	Description
Condyle	A large, rounded, articulating knob
Facet	A flattened or shallow articulating surface
Head	A prominent, rounded, articulating end of a bone
Nonarticulating prominences	
Crest	A narrow, ridgelike projection
Epicondyle	A projection above a condyle
Process	Any marked bony prominence
Spine	A sharp, slender process
Trochanter	A massive process found only on the femur
Tubercle	A small rounded process
Tuberosity	A large roughened process
Depressions and openings	
Alveolus	A deep pit or socket
Fissure	A narrow, slitlike opening
Foramen (plural, *foramina*)	A rounded opening through a bone
Fossa	A flattened or shallow surface
Fovea	A small pit or depression; some are articular and some are not
Meatus, or canal	A tubelike passageway through a bone
Sinus	A cavity or hollow space in a bone
Sulcus	A groove in a bone

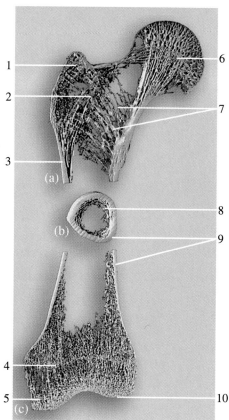

Figure 6.1 Sections of a human femur showing the outer compact bone and inner spongy bone. Note the trajectorial (stress) lines of the trabeculae of spongy bone. (a) A coronal section of the proximal end of the femur, (b) a cross section of the body of the femur, and (c) a coronal section of the distal end of the femur.

1. Greater trochanter
2. Spongy bone
3. Compact bone
4. Epiphyseal line (reminant of epiphyseal plate)
5. Femoral epicondyle
6. Head of femur
7. Trabeculae of spongy bone
8. Spongy bone surrounding medullary cavity
9. Compact bone
10. Condyle of femur

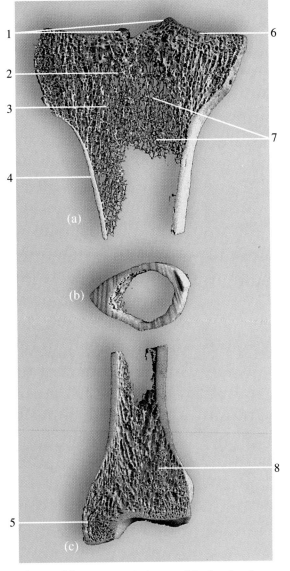

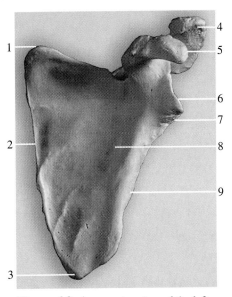

Figure 6.3 An anterior view of the left scapula.

1. Superior angle	6. Glenoid fossa
2. Medial (vertebral) border	7. Infraglenoid tubercle
3. Inferior angle	8. Subscapular fossa
4. Acromion	9. Lateral (axillary) border
5. Coracoid process	

Figure 6.2 Sections of a human tibia showing the outer compact bone and the inner spongy bone. Note the trajectorial stress lines of the trabeculae of spongy bone. (a) A coronal section of the proximal end of the tibia, (b) a cross section of the body of tibia, and (c) a coronal section of the distal end of the tibia.

1. Intercondylar eminences	5. Medial malleolus
2. Epiphyseal line	6. Lateral condyle
3. Spongy bone	7. Trabeculae of spongy bone
4. Compact bone	8. Spongy bone

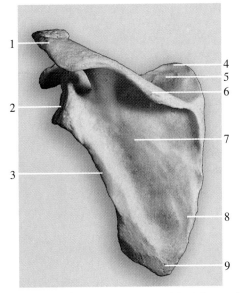

Figure 6.4 A posterior view of the left scapula.

1. Acromion	6. Spine
2. Glenoid fossa	7. Infraspinous fossa
3. Lateral (axillary) border	8. Medial (vertebral) border
4. Superior angle	9. Inferior angle
5. Supraspinous fossa	

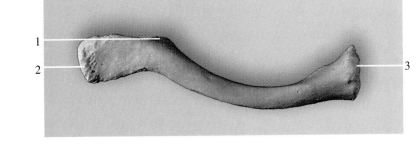

Figure 6.5 An inferior view of the clavicle.
1. Conoid tubercle
2. Acromial extremity
3. Sternal extremity

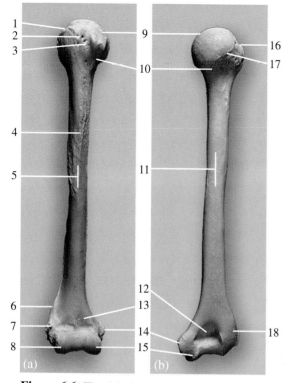

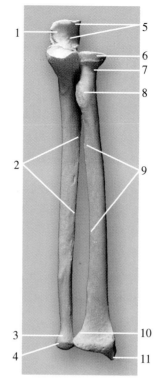

Figure 6.6 The right humerus. (a) An anterior view and (b) a posterior view.

1. Greater tubercle
2. Intertubercular groove
3. Lesser tubercle
4. Deltoid tuberosity
5. Anterior surface of humerus
6. Lateral supracondylar ridge
7. Lateral epicondyle
8. Capitulum
9. Head of humerus
10. Surgical neck
11. Posterior surface of humerus
12. Olecranon fossa
13. Coronoid fossa
14. Medial epicondyle
15. Trochlea
16. Greater tubercle
17. Anatomical neck
18. Lateral epicondyle

Figure 6.7 An anterior view of the left ulna and radius.

1. Olecranon
2. Interosseous margin
3. Head of ulna
4. Styloid process of ulna
5. Trochlear notch
6. Head of radius
7. Neck of radius
8. Radial tuberosity
9. Interosseous margin
10. Ulnar notch of radius
11. Styloid process of radius

Figure 6.8 A posterior view of the left ulna and radius.

1. Head of radius
2. Neck of radius
3. Interosseous margin
4. Styloid process of radius
5. Olecranon
6. Radial notch of ulna
7. Interosseous margin
8. Head of ulna
9. Styloid process of ulna

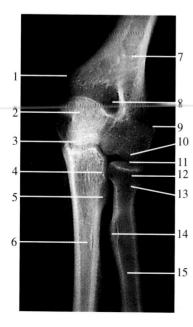

Figure 6.9 A radiograph of the left elbow region, posterior view.

1. Medial epicondyle of humerus
2. Olecranon
3. Trochlea
4. Radial notch of ulna
5. Tuberosity of ulna
6. Ulna
7. Humerus
8. Olecranon fossa of humerus
9. Lateral epicondyle of humerus
10. Capitulum
11. Articular surface of radius
12. Head of radius
13. Neck of radius
14. Tuberosity of radius
15. Radius

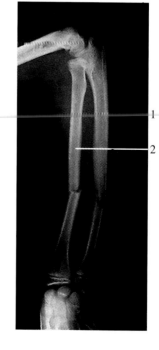

Figure 6.10 A radiograph showing fractures of the ulna and radius of a ten-year-old child. Notice the distal epiphyseal plates of the ulna and the radius.

1. Ulna
2. Radius

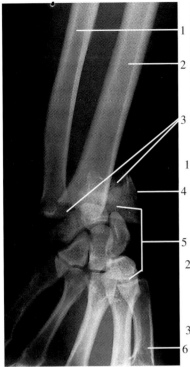

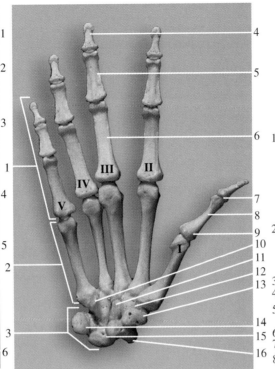

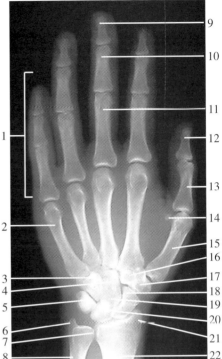

Figure 6.11 A radiograph showing a fracture of the distal portion of the radius and medial displacement of the antebrachium.
1. Ulna
2. Radius
3. Site of fracture
4. Styloid process of radius
5. Carpal bones
6. First metacarpal bone

Figure 6.12 A posterior view of the left wrist and hand.
1. Phalanges of fifth digit
2. Metacarpal bones (first-fifth)
3. Carpal bones
4. Distal phalanx of third digit
5. Middle phalanx of third digit
6. Proximal phalanx of third digit
7. Head of phalanx
8. Body of phalanx
9. Base of phalanx
10. Hamate bone
11. Capitate bone
12. Trapezoid bone
13. Trapezium bone
14. Triquetrum bone
15. Lunate bone
16. Scaphoid bone

Figure 6.13 A radiograph of the right wrist and hand, anteroposterior view.
1. Phalanges of fifth digit
2. Fifth metacarpal bone
3. Hamate bone
4. Capitate bone
5. Triquetrum and pisiform bones
6. Styloid process of ulna
7. Distal radioulnar joint
8. Ulna
9. Distal phalanx of third digit
10. Middle phalanx of third digit
11. Proximal phalanx of third digit
12. Distal phalanx of pollex (thumb)
13. Proximal phalanx of pollex (thumb)
14. Sesamoid bone
15. First metacarpal bone
16. Trapezoid bone
17. Trapezium bone
18. Scaphoid bone
19. Capitate bone
20. Lunate bone
21. Styloid process of radius
22. Radius

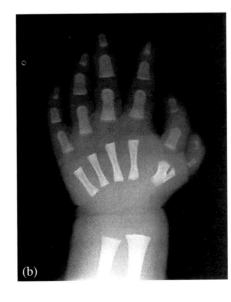

Figure 6.14 A photograph (a) and a radiograph (b) showing polydactyly, having extra digits. Polydactyly is a common congenital deformity of the hand, although it also occurs in the foot. Notice that the carpal bones are still cartilaginous in a newborn and do not show up on a radiograph.

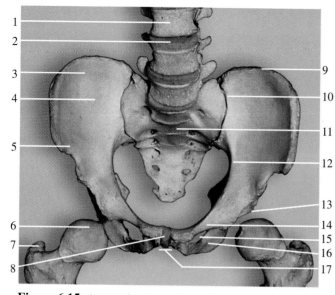

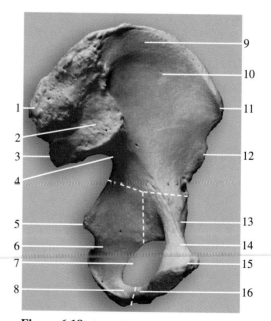

Figure 6.15 An anterior view of the articulated pelvic girdle showing the two coxal bones, the sacrum, and the two femora.

1. Lumbar vertebra	6. Head of femur	12. Pelvic brim
2. Intervertebral disc	7. Greater trochanter	13. Acetabulum
3. Ilium	8. Symphysis pubis	14. Pubic crest
4. Iliac fossa	9. Crest of the ilium	15. Obturator foramen
5. Anterior superior iliac spine	10. Sacroiliac joint	16. Ischium
	11. Sacrum	17. Pubic angle

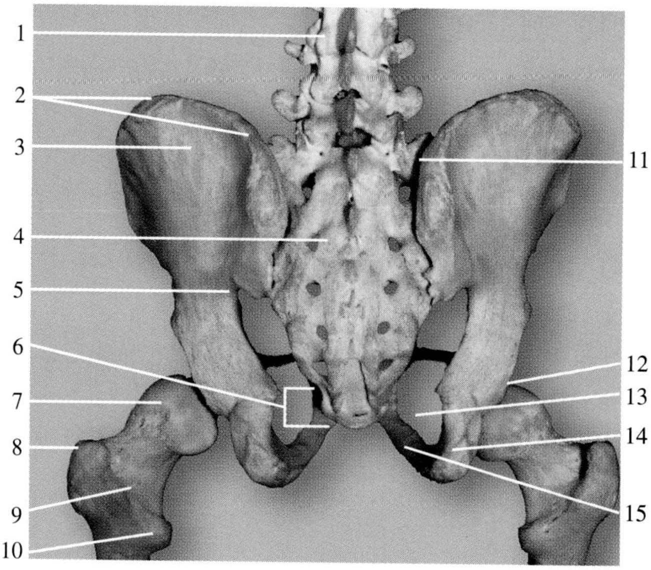

Figure 6.16 A posterior view of the articulated pelvic girdle showing the two coxal bones, the sacrum, and the two femora.

1. Lumbar vertebra	7. Head of femur	11. Sacroiliac joint
2. Crest of ilium	8. Greater trochanter	12. Acetabulum
3. Ilium	9. Intertrochanteric crest	13. Obturator foramen
4. Sacrum	10. Lesser trochanter	14. Ischium
5. Greater sciatic notch		15. Pubis
6. Coccyx		

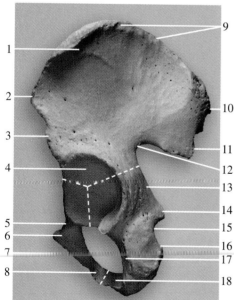

Figure 6.17 A lateral view of the left os coxae.

1. Ilium	10. Posterior superior iliac spine
2. Anterior superior iliac spine	11. Posterior inferior iliac spine
3. Anterior inferior iliac spine	12. Greater sciatic notch
4. Acetabulum	13. Ischial body
5. Superior ramus of pubis	14. Ischial spine
6. Pubis	15. Lesser sciatic notch
7. Obturator foramen	16. Ischial tuberosity
8. Inferior ramus of pubis	17. Ischium
9. Crest of the ilium	18. Ischial ramus

Figure 6.18 A medial view of the left os coxae.

1. Posterior superior iliac spine	9. Ilium
2. Auricular surface	10. Iliac fossa
3. Posterior inferior iliac spine	11. Anterior superior iliac spine
4. Greater sciatic notch	12. Anterior inferior iliac spine
5. Ischial spine	13. Superior ramus of pubis
6. Ischium	14. Pubis
7. Obturator foramen	15. Pubic crest
8. Ischial ramus	16. Pubic ramus

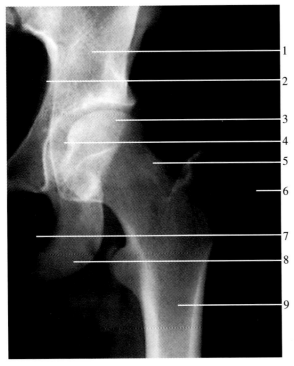

Figure 6.19 A radiograph of the left os coxae and femur, anterior view.

1. Ilium
2. Pelvic brim
3. Head of femur
4. Acetabulum
5. Neck of femur

6. Greater trochanter
7. Obturator foramen
8. Ischium
9. Femur

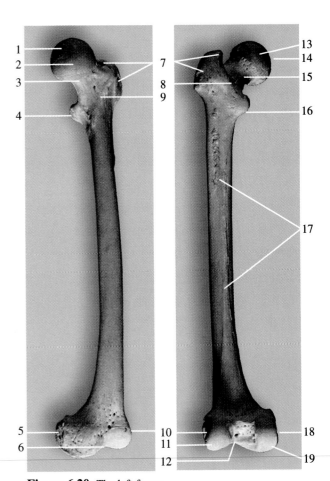

Figure 6.20 The left femur.

1. Fovea capitis femoris
2. Head
3. Neck
4. Lesser trochanter
5. Medial epicondyle
6. Patellar surface
7. Greater trochanter
8. Intertrochanteric crest
9. Intertrochanteric line
10. Lateral epicondyle

11. Lateral condyle
12. Intercondylar fossa
13. Head
14. Fovea capitis femoris
15. Neck
16. Lesser trochanter
17. Linea aspera on shaft
    (body) of femur
18. Medial epicondyle
19. Medial condyle

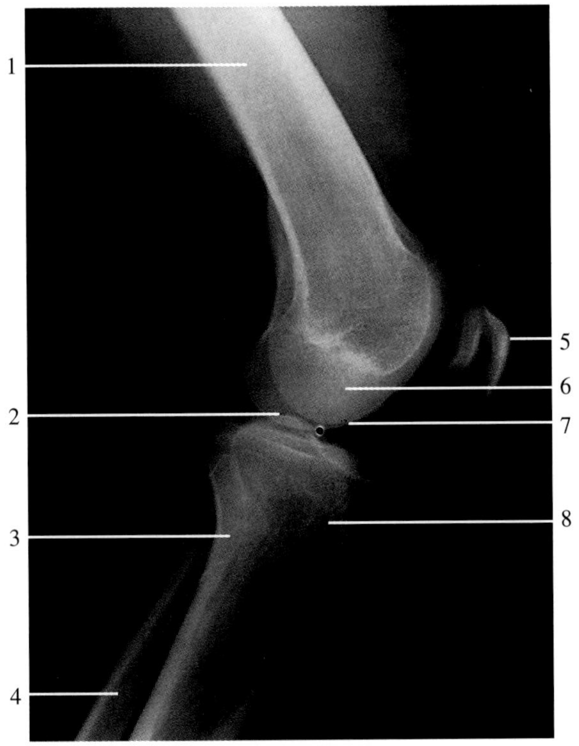

Figure 6.21 Radiograph of the left knee region, lateral view.

1. Femur
2. Intercondylar eminences of tibia
3. Tibia
4. Fibula
5. Patella
6. Lateral epicondyle of femur
7. Condyle of femur
8. Tibial tuberosity

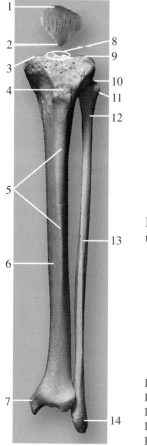

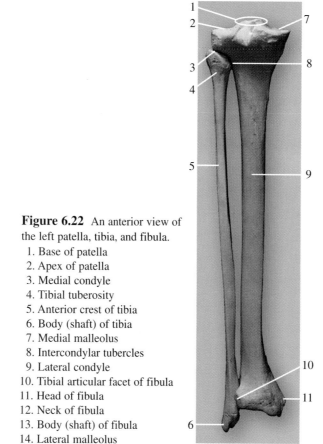

Figure 6.22 An anterior view of the left patella, tibia, and fibula.
1. Base of patella
2. Apex of patella
3. Medial condyle
4. Tibial tuberosity
5. Anterior crest of tibia
6. Body (shaft) of tibia
7. Medial malleolus
8. Intercondylar tubercles
9. Lateral condyle
10. Tibial articular facet of fibula
11. Head of fibula
12. Neck of fibula
13. Body (shaft) of fibula
14. Lateral malleolus

Figure 6.23 A posterior view of the left tibia and fibula.
1. Intercondylar tubercles
2. Lateral condyle of tibia
3. Head of fibula
4. Neck of fibula
5. Body (shaft) of fibula
6. Lateral malleolus
7. Medial condyle of tibia
8. Fibular articular facet of tibia
9. Body (shaft) of tibia
10. Fibular notch of tibia
11. Medial malleolus

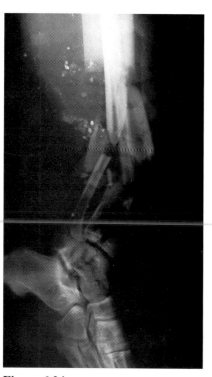

Figure 6.24 A severe fracture of the leg and ankle (talus bone). In this patient, the trauma was so extensive amputation of the leg was necessary.

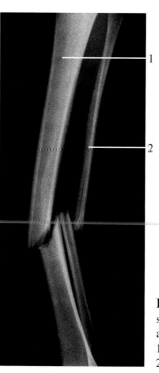

Figure 6.25 Radiograph showing fractures of the tibia and fibula.
1. Tibia
2. Fibula

Figure 6.26 An inferior view of the left foot.

1. Phalanges of fifth digit
2. Metatarsal bones (first–fifth)
3. Tarsal bones
4. Distal phalanx of first digit
5. Proximal phalanx of first digit
6. Head of first metatarsal bone
7. Body of first metatarsal bone
8. Base of first metatarsal bone
9. Medial cuneiform
10. Intermediate cuneiform
11. Lateral cuneiform
12. Navicular bone
13. Cuboid bone
14. Talus bone
15. Calcaneus bone
16. Tuberosity of calcaneus

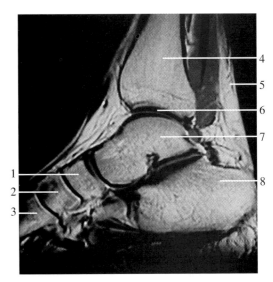

Figure 6.27 An MR image of the ankle region, medial view.

1. Navicular bone	5. Tendo calcaneus
2. Medial cuneiform	6. Tibiotalar joint
3. First metatarsal bone	7. Talus
4. Tibia	8. Calcaneus

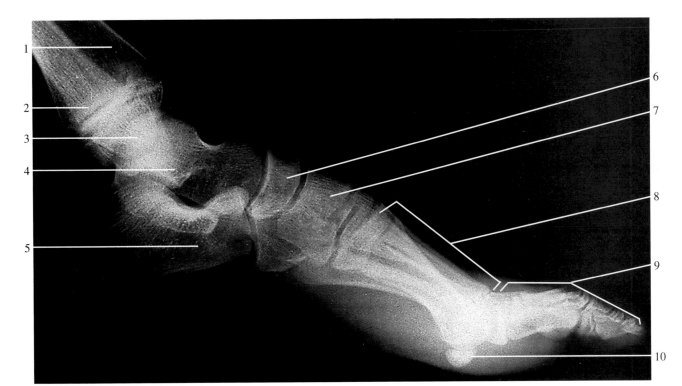

Figure 6.28 A radiograph of the left foot, medial view.

1. Tibia	3. Tibiotalar joint	5. Calcaneus	7. First cuneiform	9. Phalanges
2. Fibula (superimposed)	4. Talus	6. Navicular bone	8. Metatarsal bones	10. Sesamoid bone

Articulations

7

The structure of a joint determines its range of movement. Not all joints are flexible, however, and as one part of the body moves, other joints remain rigid to stabilize the body and maintain balance. *Arthrology* is the study of joints and *kinesiology* is the study of body movement, or the functional relationship between the skeleton, joints, muscles, and innervation (nerve supply) as they work together to produce coordinated movement.

The joints of the body are structurally classified into three basic kinds:

1. **Fibrous** – fibrous connective tisues join the skeletal structures; the joints lack a joint capsule, but in some, slight movement is possible.

2. **Cartilaginous** – fibrocartilage or hyaline cartilage joins the skeletal structures; the joints lack a joint capsule, but in some, slight movement is possible.

3. **Synovial** – joint capsules containing synovial fluid are present between the articulating bones; articular cartilages and ligaments supporting the articulating bones are also present, which permit freedom of movement. Synovial joints are the freely moveable joints of the body.

Table 7.1 Movements Permitted at Synovial Joints

Type of movement	Description
Angular movement	Increase or decrease the joint angle
Flexion	Decreasing the angle between two bones
Extension	Increasing the angle between two bones
Hyperextension	Excessive extension beyond 180° (angle of anatomical position)
Dorsiflexion	Bending the foot toward the tibia
Plantar flexion	Bending the foot away from the tibia
Abduction	Movement of a body part away from the axis of the body, or away from the midsagittal plane, in a lateral direction
Inversion	Movement of the sole of the foot inward, or medially
Eversion	Movement of the sole of the foot outward, or laterally
Circular movement	
Rotation	Turning of a bone on its own axis
Circumduction	Movement of a body segment in a circular, conelike motion

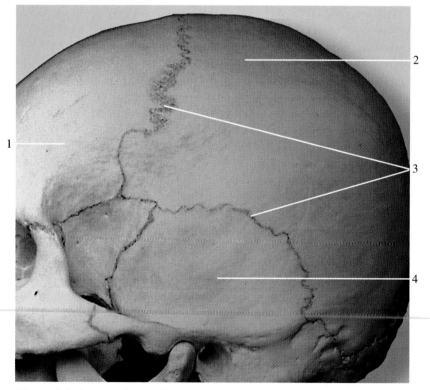

Figure 7.1 A suture located between cranial bones is a type of fibrous joint.
1. Frontal bone
2. Parietal bone
3. Sutures
4. Temporal bone

Table 7.2 Types of Joints

Type	Structure and movement	Example
Fibrous	Fibrous connective tissue joins the skeletal structures	
Suture	Frequently serrated edges of articulating bones, separated by a thin layer of fibrous tissue; no movement	Sutures between bones, cranial bones
Syndesmoses	Articulating bones bound by an interosseous ligament; slight movement	Joints between tibia-fibula and radius-ulna
Gomphoses	Periodontal ligament binding teeth into aveoli of bone; no movement	Teeth secured in the dental alveoli (teeth sockets)
Cartilaginous	Fibrocartilage or hyaline cartilage joins the skeletal structures	
Symphyses	Thin pad of fibrocartilage between articulating bones; slight movement	Symphysis pubis, sacroiliac joint and intervertebral joints
Synchondroses	Mitotically active hyaline cartilage between skeletal structures; no movement	Epiphyseal plates between diaphysis and epiphyses of long bones
Synovial	Joint capsule between articulating bones, containing synovial fluid; extensive movement	
Gliding	Flattened or slightly curved articulating surfaces; sliding movement	Intercarpal and intertarsal joints
Hinge	Concave surface of one bone articulates with a depression of another; bending motion in one plane	Humeroulnar (elbow) and tibiofemoral (knee) joints; interphalangeal (finger) joints of digits
Pivot	Conical surface of one bone articulates with a depression of another; rotation about a central axis; rotational movement	Atlantoaxial joint; proximal radioulnar joint
Condyloid	Oval condyle of one bone articulates with elliptical cavity of another; biaxial movement	Radiocarpal (wrist) joint
Saddle	Concave and convex surface on each articulating bone; wide range of movement	Carpometacarpal joint at the base of the thumb
Ball-and-socket	Rounded convex surface of one bone articulates with cuplike socket of another; movement in all planes and rotation	Glenohumeral (shoulder) and coxal (hip) joints

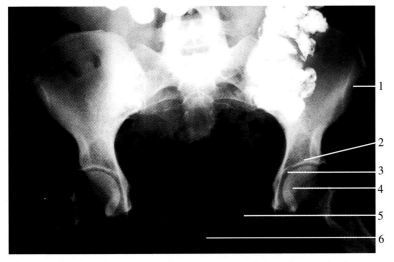

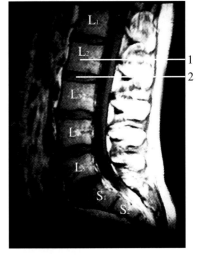

Figure 7.2 A symphysis is a type of cartilaginous joint. The symphysis pubis can be readily seen in this radiograph of the pelvic girdle. Note the head of the femur articulating with the acetabulum of the os coxae forming a ball-and-socket joint. A ball-and-socket joint is a freely moveable synovial joint.

1. Ilium
2. Acetabulum
3. Ball-and-socket joint
4. Head of femur
5. Pubis
6. Symphysis pubis

Figure 7.3 An intervertebral joint between vertebral bodies is a type of cartilaginous joint. The intervertebral discs of intervertebral joints can be readily seen in a lateral MR image of the lower back.

1. Body of second lumbar vertebra (L2)
2. Intervertebral disc

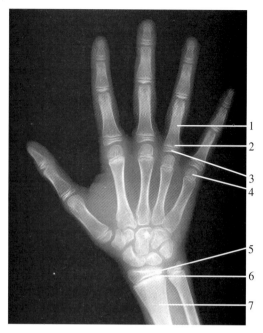

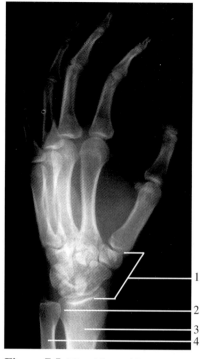

Figure 7.4 A synchondrosis is a type of cartilaginous joint located within a long bone at both the proximal and the distal epiphyseal plates as seen in a radiograph of a child's hand. Mitotic activity at a synchondrotic joint is responsible for linear (length) bone growth.

1. Diaphysis of phalanyx
2. Epiphyseal plate
3. Proximal epiphysis of phalanx
4. Distal epiphysis of metacarpal bone
5. Distal epiphysis of radius
6. Epiphyseal plate
7. Diaphysis of radius

Figure 7.5 The side-to-side articulation of the ulna and radius forms a syndesmosis that is tightly bound by an interosseous ligament (not seen in radiograph). A syndesmosis is a type of fibrous joint that permits slight movement.

1. Carpal bones   3. Radius
2. Syndesmosis   4. Ulna

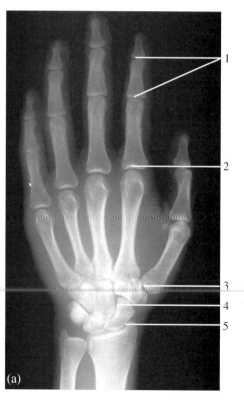

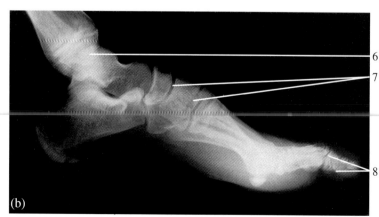

Figure 7.6 Radiographs of the (a) wrist and hand and (b) the ankle and foot. Several kinds of synovial joints (freely moveable joints) are contained in these regions.

1. Hinge joints — metacarpophalangeal joints of second digit (index finger)
2. Condyloid joint — metacarpophalangeal joint of the second digit (index finger)
3. Saddle joint — carpometacarpal joint of the pollex (thumb)
4. Gliding joint — intercarpal joint between capitate and scaphoid bones
5. Condyloid joint — radiocarpal joint
6. Condyloid joint at ankle
7. Gliding joints between tarsal bones
8. Hinge joints between phalanges of digits

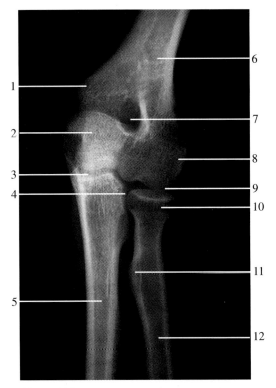

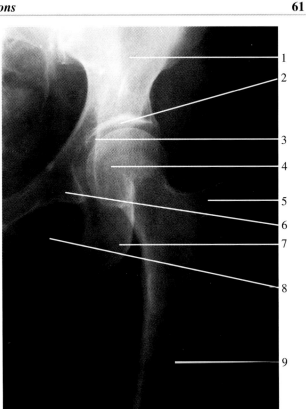

Figure 7.7 A radiograph of the elbow region depicting two types of synovial joints. The humeroulnar joint is a hinge joint that permits movement along a single plane. The proximal radioulnar and humeroradial joints are both pivot joints that permit rotational movement.

1. Medial epicondyle of humerus
2. Olecranon
3. Humeroulnar joint
4. Proximal radioulnar joint
5. Ulna
6. Humerus
7. Olecranon fossa
8. Lateral epicondyle of humerus
9. Humeroradial joint
10. Head of radius
11. Radial tuberosity
12. Radius

Figure 7.8 The ball-and-socket joint seen in this radiograph of the hip is a synovial joint. The coxal joint (hip joint) is formed by the head of the femur articulating with the acetabulum of the os coxae.

1. Ilium
2. Acetabulum
3. Ball-and-socket joint
4. Head of femur
5. Greater trochanter of femur
6. Pubis
7. Ischium
8. Obturator foramen
9. Femur

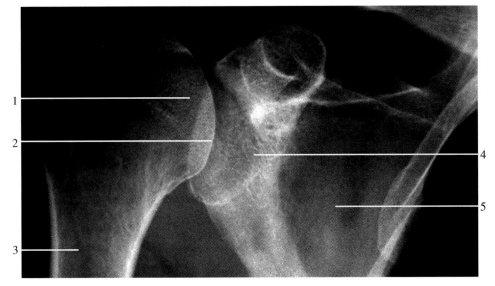

Figure 7.9 The ball-and-socket joint seen in this radiograph of the shoulder is a synovial joint. The glenohumeral joint (shoulder joint) is formed by the head of the humerus articulating with the glenoid fossa of the scapula.

1. Head of humerus
2. Ball-and-socket joint
3. Humerus
4. Glenoid fossa
5. Scapula

(a)

Flexion at right shoulder,
elbow, and knee joints.
Extension at left shoulder,
elbow, and knee joints.

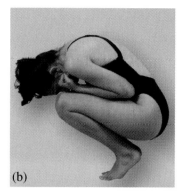

Maximal flexion at each
of the principal body
joints.

(b)

(c)

Rotation at the joints of
the neck; elevation at
the shoulder joint; and
flexion at the elbow
and wrist joints.

(d)

Adduction at the shoulder
and hip joints. Also,
adduction at the joints of
the fingers and toes.

(e)

Abduction at the shoulder and
hip joints. Also, abduction at the
metacarpophalangeal joints at the
base of the digits.

(f)

Rotation at joints of the
vertebral column.

(g)

Lateral bending at joints of
the vertebral column.

(h)

Flexion at the joints of the
vertebral column.

(i)

Hyperextension at the joints
of the vertebral column.

(j)

Flexion at the shoulder, hip,
and knee joints on left side of
body; extension at the elbow
and wrist joints.

(k)

Extension at the shoulder
and hip joints on left side of
body; plantar flexion at the
left ankle joint.

Figure 7.10 A photographic summary of joint movements (a-k).

(a)

Flexion, extension, and hyperextension at the cervical intervertebral joints of the neck.

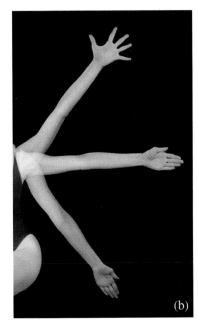

(b)

Adduction and abduction of the right arm at the shouder joint and fingers at the metacarpophalangeal joints.

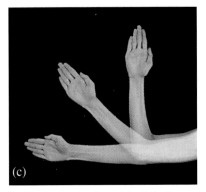

(c)

Flexion and extension at the left elbow joint.

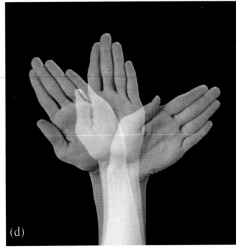

(d)

Abduction and adduction of the hand at the wrist.

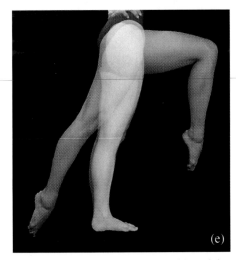

(e)

Flexion and extension at the hip and knee joints.

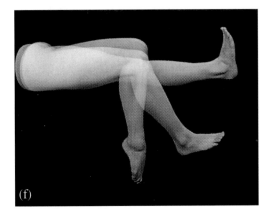

(f)

Flexion and extension at the knee joint and plantar flexion and dorsiflexion at the ankle joint.

Figure 7.11 A visualization of angular movements permitted at synovial joints (a-f).

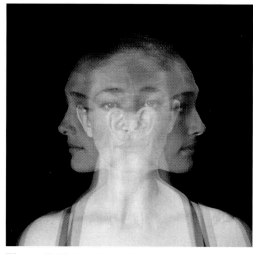

Figure 7.12 Rotation of the head at the cervical vertebrae–principally at the atlantoaxial joint.

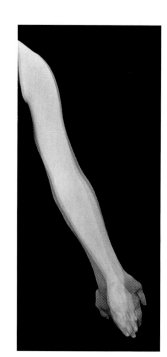

Figure 7.13 Rotation of the antebrachium (forearm) at the proximal radioulnar joint––an example of pronation and supination.

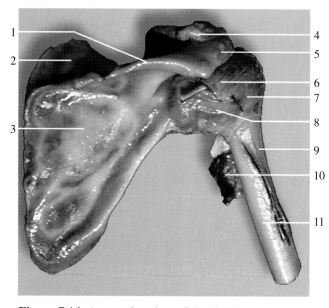

Figure 7.14 A posterior view of the right scapula, proximal humerus, and glenohumeral (shoulder) joint.

1. Spine of Scapula
2. Supraspinous fossa
3. Infraspinous fossa
4. Coracoacrominal ligament
5. Acromion
6. Tendon of infraspinatus muscle
7. Tendon of teres minor muscle
8. Articular capsule of glenohumeral joint
9. Origin of lateral head of triceps brachii muscle
10. Teres major muscle
11. Humerus

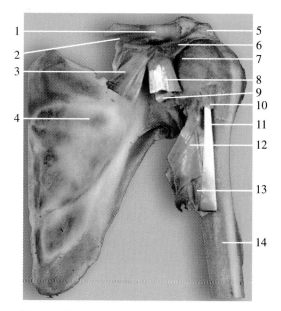

Figure 7.15 An anterior view of the right scapula, proximal humerus, and gleno-humeral (shoulder) joint.

1. Clavicle
2. Subclavius muscle
3. Tendon of pectoralis minor muscle
4. Subscapular fossa
5. Acromion
6. Coracoid process of scapula
7. Articular capsule of glenohumeral joint
8. Tendon of short head of biceps brachii muscle
9. Tendon of coracobrachialis muscle
10. Tendon sheath
11. Tendon of long head of biceps brachii muscle
12. Tendon of latissimus dorsi muscle
13. Tendon of teres major muscle
14. Humerus

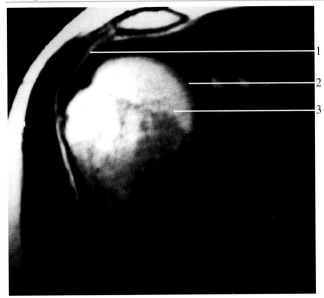

Figure 7.16 Structure of the glenohumeral joint (shoulder joint) as seen in a MR image.
1. Glenoid labrum
2. Articular cavity of shoulder joint
3. Head of humerus

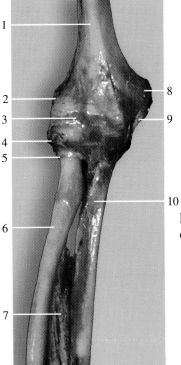

Figure 7.17 An anterior view of the right elbow region.
1. Humerus
2. Lateral epicondyle of humerus
3. Articular capsule of elbow joint
4. Annular ligament
5. Head of radius
6. Radius
7. Interosseous ligament
8. Medial epicondyle
9. Ulnar collateral ligament
10. Ulna

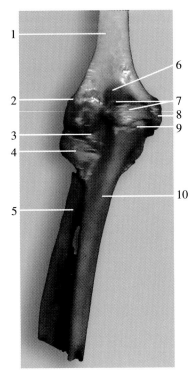

Figure 7.18 A posterior view of the right elbow region.

1. Humerus
2. Medial epicondyle
3. Coronoid process of ulna
4. Head of radius
5. Radius
6. Olecranon fossa
7. Articular capsule of elbow joint
8. Lateral epicondyle
9. Olecranon
10. Ulna

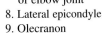
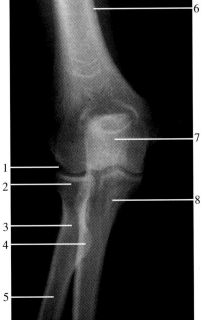

Figure 7.19 Structure of the elbow joint as seen in a radiograph.
1. Articular cavity of elbow joint
2. Head of radius
3. Neck of radius
4. Radial tuberosity
5. Radius
6. Humerus
7. Olecranon
8. Ulna

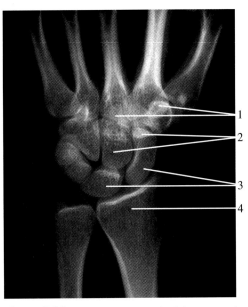

Figure 7.20 Structure of the wrist joint as seen in a MR image.
1. Heads of second and third metacarpal bones
2. Distal carpal bones (trapezoid and capitate bones)
3. Proximal carpal bones (scaphoid and lunate bones)
4. Distal epiphysis of radius

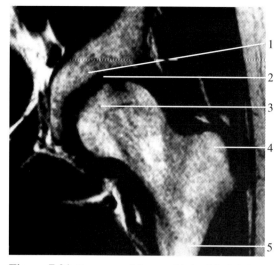

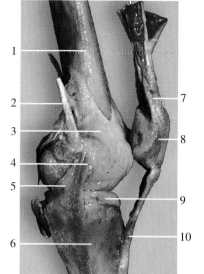

Figure 7.22 A medial view of right tibiofemoral (knee) joint
1. Femur
2. Tendon of adductor magnus muscle
3. Articular capsule of tibiofemoral (knee) koint
4. Medial epicondyle of femur
5. Tibial collateral ligament
6. Tibia
7. Tendon of quadriceps femoris muscle
8. Patella
9. Articular cartilage of tibia
10. Patellar ligament

Figure 7.21 Structure of the coxal joint (hip joint) as seen in a MR image.
1. Acetabulum
2. Articular cavity of hip joint
3. Head of femur
4. Greater trochanter of femur
5. Body of femur

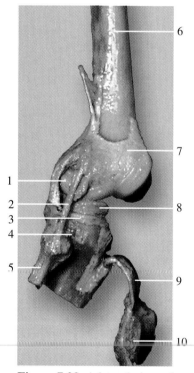

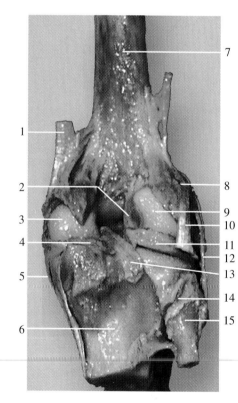

Figure 7.24 A posterior view of right tibiofemoral (knee) joint.
1. Tendon of adductor magnus muscle
2. Anterior cruciate ligament
3. Medial femoral condyle
4. Medial meniscus
5. Medial collateral ligament
6. Tibia
7. Femur
8. Lateral epicondyle of femur
9. Lateral femoral condyle
10. Tendon of popliteus muscle
11. Lateral meniscus
12. Lateral collateral ligament
13. Posterior cruciate ligament
14. Proximal tibiofibular joint
15. Head of fibula

Figure 7.23 A lateral view of right tibiofemoral (knee) joint.
1. Bursa
2. Lateral collateral ligament
3. Articular cartilage
4. Proximal
5. Tibiofibular joint
6. Femur
7. Articular capsule
8. Lateral meniscus
9. Patellar ligament
10. Patella

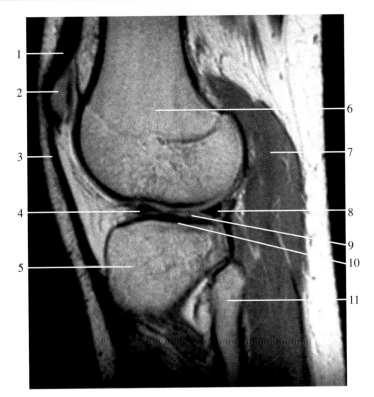

Figure 7.25 A sagittal MR image of tibiofemoral joint (knee joint).

1. Tendon of quadriceps muscle
2. Patella
3. Patellar ligament
4. Lateral meniscus, anterior horn
5. Head of tibia
6. Femur
7. Gastrocnemius muscle
8. Lateral meniscus, posterior horn
9. Synovial fluid
10. Articular cartilage
11. Head of fibula

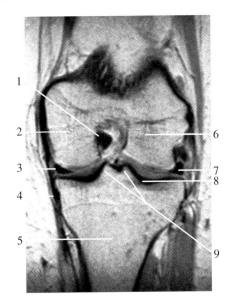

Figure 7.26 A posterior MR image of tibiofemoral joint (knee joint).

1. Insertion of posterior cruciate ligament
2. Medial condyle
3. Medial meniscus
4. Tibial collateral ligament
5. Tibia
6. Lateral condyle of femur
7. Lateral meniscus
8. Synovial fluid
9. Intercondylar eminences

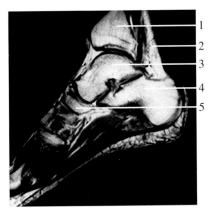

Figure 7.27 A sagittal MR image of the talocrual (ankle) and tarsal joints of the foot.

1. Tibia
2. Tendo calcaneus
3. Talus
4. Calcaneus
5. Navicular bone

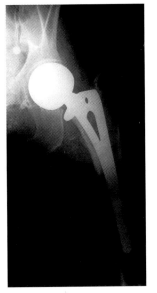

Figure 7.28 A radiograph of the hip showing a joint prosthesis.

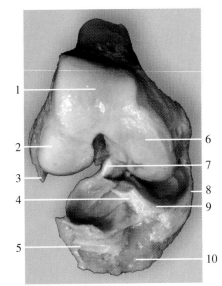

Figure 7.29 An anterior view of right tibiofemoral (knee) joint as it is flexed.

1. Patellar articular surface
2. Lateral femoral condyle
3. Lateral collateral ligament
4. Anterior cruciate ligament
5. Articular cartilage of tibia
6. Medial femoral condyle
7. Posterior cruciate ligament
8. Medial collateral ligament
9. Medial meniscus
10. Tibia

Muscular System

Most skeletal muscles span joints and are attached to a bone at both ends by a *tendon*. Certain tendons, especially at the wrist and ankle, are enclosed individually by *tendon sheaths*. A group of tendons in these areas is also covered by a *retinaculum*. The *origin* of a muscle is the more stationary attachment, and the *insertion* is the more moveable attachment (fig. 8.1). *Synergistic muscles* contract together. *Antagonistic muscles* perform in opposition to a synergistic group of muscles.

Fascia is a fibrous connective tissue that covers muscle and attaches to the skin. *Superficial fascia* secures the skin to the underlying muscle. *Deep fascia* is an inward extension of the superficial fascia and surrounds adjacent muscles compartmentalizing and binding them into functional groups.

A skeletal muscle fiber is an elongated, multinucleated, striated cell (fig. 8.2). Each fiber is surrounded by a cell membrane, called a *sarcolemma*, and the cytoplasm within the cell is called *sarcoplasm*. Each fiber contains many parallel, thread-like structures called *myofibrils*. Each myofibril is composed of smaller strands called *myofilaments* that contain the contractile proteins, *actin* and *myosin*. The regular spatial organization of the contractile proteins within the myofibrils forms the cross-banding striations characteristic of skeletal muscle. A network of membranous channels, called the *sarcoplasmic reticulum*, extends throughout the cytoplasm.

The myofibrils of a skeletal muscle fiber are arranged into compartments called *sarcomeres*. The dark and light striations of the sarcomere are due to the arrangement of the thick (myosin) and thin (actin) filaments. The dark bands are called *A bands* and the lighter bands, containing only actin filaments, are called *I bands*. The outer regions of the A bands contain actin and myosin; however, the lighter central regions (*H zones*) of the A bands contain only myosin. The I bands are bisected by dark *Z lines* where the actin filaments of adjacent sarcomeres join. Muscle fiber contraction results from the interaction of contractile proteins (actin and myosin myofilaments) in which the length of the sarcomeres is reduced.

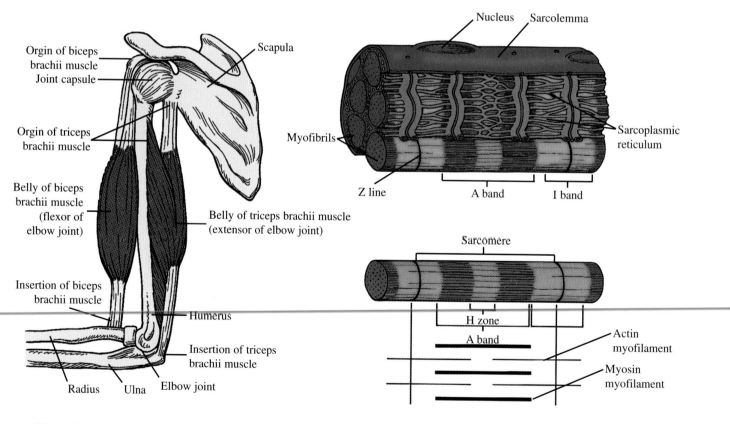

Figure 8.1 Skeletomuscular relationship.

Figure 8.2 Structure of a muscle fiber.

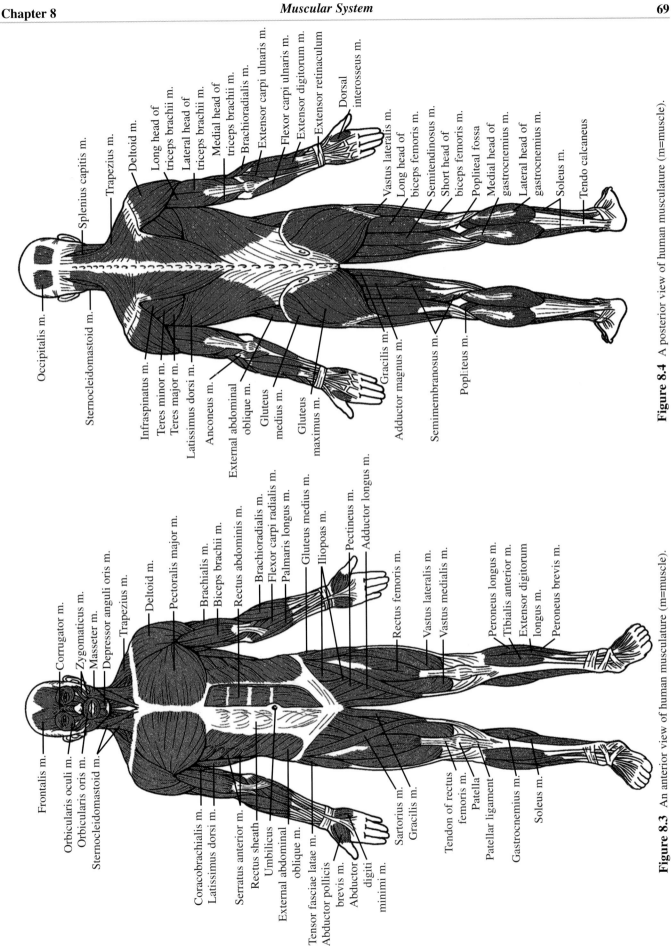

Figure 8.4 A posterior view of human musculature (m=muscle).

Figure 8.3 An anterior view of human musculature (m=muscle).

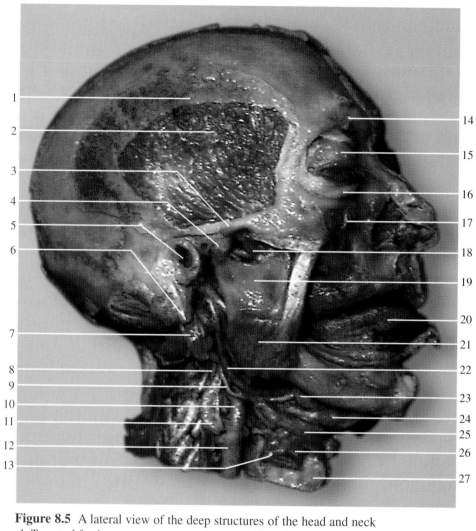

Figure 8.5 A lateral view of the deep structures of the head and neck

1. Temporal fascia
2. Temporalis muscle
3. Zygomatic arch
4. Joint capsule of temporomandibular joint
5. External acoustic canal
6. Mastoid process of temporal bone
7. Posterior belly of digastric muscle
8. Vagus nerve
9. Hypoglossal nerve
10. Internal cartoid artery
11. External cartoid artery
12. Common carotid artery
13. Omohyoid muscle
14. Supraorbital nerve
15. Superior tarsal plate
16. Palpebral fascia
17. Infraorbital nerve
18. Lateral pterygoid muscle
19. Mandible
20. Tongue
21. Masseter muscle
22. Stylohyoid muscle
23. Mandibular gland
24. Anterior belly of digastric muscle
25. Mylohyoid muscle
26. Sternohyoid muscle
27. Thyroid cartilage of larynx

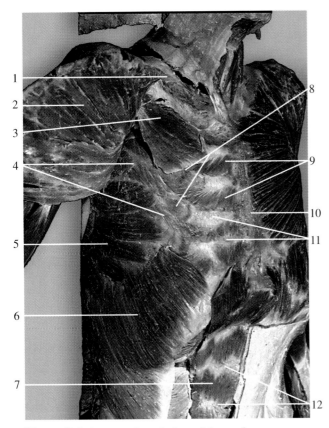

Figure 8.6 An anterolateral view of the trunk.

1. Clavicle
2. Pectoralis major m. (reflected)
3. Pectoralis minor m.
4. External intercostal m.
5. Serratus anterior m.
6. External abdominal oblique m.
7. Rectus abdominis m.
8. Ribs
9. Internal intercostal m.
10. Sternum
11. Costal cartilages
12. Tendinous inscriptions of rectus abdominis m.

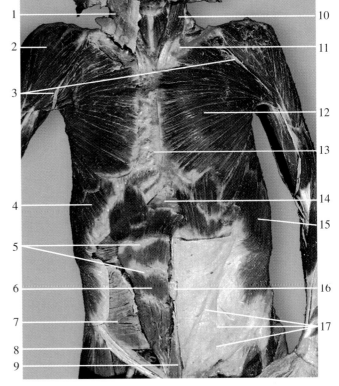

Figure 8.7 An anterior view of the trunk.

1. Platysma m.
2. Deltoid m.
3. Cephalic veins
4. External abdominal oblique m. (aponeurosis removed)
5. Tendinous inscriptions of rectus abdominis m.
6. Rectus abdominis m.
7. Internal abdominal oblique m.
8. Inguinal ligament
9. Pyramidalis m.
10. Sternocleidomastoid m.
11. Clavicle
12. Pectoralis major m.
13. Sternum
14. Xiphoid process
15. External abdominal oblique m.
16. Umbilicus
17. Aponeurosis of external abdominal oblique m.

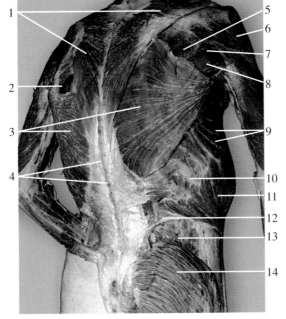

Figure 8.8 A posterolateral view of the trunk.

1. Trapezius m.
2. Triangle of ausculation
3. Latissimus dorsi m.
4. Vertebral column (spinous processes)
5. Infraspinatus m.
6. Deltoid m.
7. Teres minor m.
8. Teres major m.
9. Serratus anterior m.
10. Rib
11. External abdominal oblique m.
12. Iliac crest
13. Gluteus medius m.
14. Gluteus maximus m.

Table 8.1 Muscles of the Vertebral Column

Spinal Muscle	Origin(s)	Insertion(s)	Action
Quadratus lumborum	Iliac crest and lower three lumbar vertebrae	Twelfth rib and upper four lumbar vertebrae	Extends lumbar region; laterally flexes vertebral column
Erector spinae Iliocostalis lumborum	Crest of ilium	Lower six ribs	Extends lumbar region
Iliocostalis thoracis	Lower six ribs	Upper six ribs	Extends thoracic region
Iliocostalis cervicis	Angles of third to sixth ribs	Transverse processes of fourth to sixth cervival vertebrae	Extends cervical region
Longissimus thoracis	Transverse processes of lumbar vertebrae	Transverse processes of all the thoracic vertebrae and lower nine ribs	Extends thoracic region
Longissimus cervicis	Transverse processes of upper four or five thoracic vertebrae	Transverse processes of second to sixth cervical vertebrae	Extends cervical region and lateral flexion
Longissimus capitis	Transverse processes of upper five thoracic vertebrae	Posterior margin of cranium and mastoid process of temporal bone	Extends head; acting seperatly, turns face toward that side
Spinalis thoracis	Spinous processes of upper lumbar and lower thoracic vertebrae	Spinous processes of upper thoracic vertebrae	Extends vertebral column

Table 8.2 Muscles that Act on the Pectoral Girdle

Pectoral Muscle	Origin(s)	Insertion(s)	Action
Serratus anterior	Upper eight or nine ribs	Anterior medial border of scapula	Pulls scapula forward and upward
Pectoralis minor	Sternal ends of third, fourth, and fifth ribs	Coracoid process of scapula	Pulls scapula forward and downward
Subclavius	First rib	Subclavian groove of clavical	Draws clavicle downward
Trapezius	Occipital bone and spines of cervical and thoracic vertebrae	Clavicle, acromion and spine of scapula	Elevates, depresses, and adducts scapula; hyperextends neck; braces shoulder
Levator scapulae	First to fourth cervical vertebrae	Superior border of scapula	Elevates scapula
Rhomboideus major	Spines of second to fifth thoracic vertebrae	Medial border of scapula	Elevates and adducts scapula
Rhomboideus minor	Seventh cervical and first thoracic vertebrae	Medial border of scapula	Elevates and adducts scapula

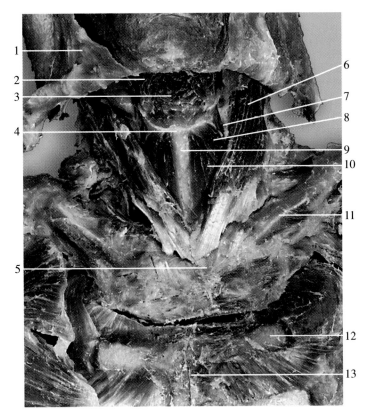

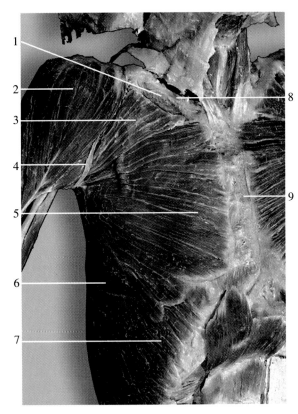

Figure 8.9 An anterior view of the neck muscles.

1. Platysma m.
2. Digastric m.
3. Mylohyoid m.
4. Hyoid bone
5. Manubrium of sternum
6. Sternocleidomastoid m.
7. Thyrohyoid m.
8. Omohyoid m.
   (superior belly)
9. Thyroid cartilage
10. Sternohyoid m.
11. Clavicle
12. Rib
13. Body of sternum

Figure 8.10 An anterolateral view of the trunk.

1. Subclavius m.
2. Deltoid m.
3. Pectoralis major m. (clavicular part)
4. Cephalic vein
5. Pectoralis major m.
6. Serratus anterior m.
7. External abdominal oblique m.
8. Clavicle
9. Sternum

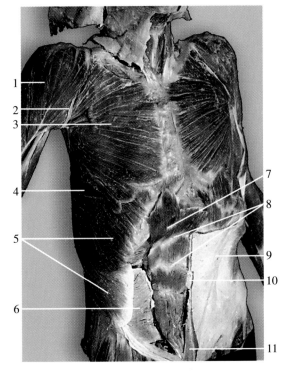

Figure 8.11 An anterolateral view of the trunk.

1. Deltoid m.
2. Cephalic vein
3. Pectoralis major m.
4. Serratus anterior m.
5. External abdominal oblique m.
6. Internal abdominal oblique m.
7. Rectus abdominis m.
8. Tendinous inscriptions
9. Aponeurosis of external abdominal oblique m.
10. Umbilicus
11. Pyramidalis m.

Table 8.3 Muscles of Mastication

Chewing Muscle	Origin(s)	Insertion(s)	Action
Temporalis	Temporal fossa	Coronoid process of mandible	Elevates mandible
Masseter	Zygomatic arch	Lateral ramus of mandible	Elevates mandible
Medial pterygoid	Sphenoid bone	Medial ramus of mandible	Depresses and laterally moves mandible
Lateral pterygoid	Sphenoid bone	Anterior side of condylar process of mandible	Protracts mandible

Table 8.4 Muscles of the Neck

Neck Muscle	Origin(s)	Insertion(s)	Action
Sternocleidomastoid	Sternum and clavicle	Mastoid process of temporal bone	Flexes neck; rotates head to side
Digastric	Inferior border of mandible and mastoid process of temporal bone	Hyoid bone	Opens mouth; elevates hyoid bone
Mylohyoid	Inferior border of mandible	Hyoid bone and median raphe	Elevates hyoid bone and floor of mouth
Geniohyoid	Medial surface of mandible at chin	Hyoid bone	Elevates hyoid bone
Stylohyoid	Styloid process of temporal bone	Hyoid bone	Elevates and retracts tongue
Sternohyoid	Manubrium	Hyoid bone	Depresses hyoid bone
Sternothyroid	Manubrium	Thyroid cartilage	Depresses thyroid cartilage
Thyrohyoid	Thyroid cartilage	Hyoid bone	Depresses hyoid bone; elevates thyroid cartilage
Omohyoid	Superior border of scapula	Hyoid bone	Depresses hyoid bone

Table 8.5 Muscles of the Abdominal Wall

Abdominal Muscle	Origin(s)	Insertion(s)	Action
External abdominal oblique	Lower eight ribs	Iliac crest and linea alba	Compresses abdomen; lateral rotation
Internal abdominal oblique	Iliac crest, lumbodorsal fascia, inguinal ligament	Linea alba and costal cartilages of last three of four ribs	Compresses abdomen; lateral rotation
Transversus abdominis	Iliac crest, lumbodorsal fascia, inguinal ligament, costal cartilages of last six ribs	Xiphoid process, linea alba, pubis	Compresses abdomen
Rectus abdominis	Pubic crest and symphysis pubis	Xiphoid process and costal cartilages of fifth to seventh ribs	Flexes vertebral column

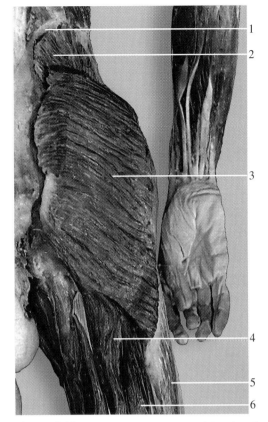

Figure 8.12 The deep muscles of the back and right side.

1. Splenius capitis m.
2. Levator scapulae m.
3. Rhomboideus minor m.
4. Rhomboideus major m.
5. Spinalis thoracis m.
6. Longissimus thoracis m.
7. Iliocostalis thoracis m.
8. Serratus posterior inferior m.
9. Iliac crest
10. Spine of the scapula
11. Supraspinatus m.
12. Deltoid m.
13. Teres minor m.
14. Infraspinatus m.
15. Teres major m.
16. Scapula (medial border)
17. Latissimus dorsi m. (reflected)
18. Serratus anterior m.
19. External intercostal m.
20. Rib
21. External abdominal oblique m.

Figure 8.13 The superficial muscles of the gluteal region.

1. Iliac crest
2. Gluteus medius m.
3. Gluteus maximus m.
4. Semitendinosus m.
5. Vastus lateralis m.
6. Biceps femoris m. (long head)

Figure 8.14 The deep structures of the gluteal region.

1. Iliac crest
2. Gluteus minimus m.
3. Piriformis m.
4. Superior gemellus m.
5. Obturator internus m.
6. Inferior gemellus m.
7. Quadratus femoris m.
8. Ischial tuberosity
9. Semimembranosus m.
10. Gluteus medius m. (reflected)
11. Gluteus maximus m. (reflected)
12. Sciatic nerve
13. Semitendinosus m.
14. Biceps femoris m. (long head)

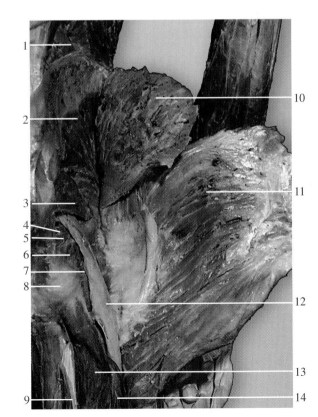

Table 8.6 Muscles that Act on the Brachium (Upper Arm)

Axial or Scapular Muscle	Origin(s)	Insertion(s)	Action
Pectoralis major	Clavicle, sternum, costal cartilages of second to sixth ribs	Greater tubercle of humerus	Flexes, adducts, and rotates shoulder joint medially
Latissimus dorsi	Spines of sacral, lumbar, and lower thoracic vertebrae; lower ribs	Intertubercular groove of humerus	Extends, adducts, and rotates shoulder joint medially; adducts shoulder joint
Deltoid	Clavicle, acromion and spine of scapula	Deltoid tuberosity of humerus	Abducts, extends, or flexes shoulder joint
Supraspinatus	Supraspinous fossa of scapula	Greater tubercle of humerus	Abducts and laterally rotates shoulder joint
Infraspinatus	Infraspinous fossa of scapula	Greater tubercle of humerus	Rotates shoulder joint laterally
Teres major	Inferior angle and lateral border of scapula	Intertubercular groove of humerus	Extends, adducts, and rotates shoulder joint medially
Teres minor	Lateral border of scapula	Greater tubercle of humerus cartilage	Rotates shoulder joint laterally
Subscapularis	Subscapular fossa	Subscapular fossa	Rotates shoulder joint medially
Coracobrachialis	Coracoid process of scapula	Body of humerus	Flexes and adducts shoulder joint

Table 8.7 Muscles that Act on the Antebrachium (Forearm)

Brachial muscle	Origin(s)	Insertion(s)	Action
Biceps brachii	Coracoid process and tuberosity above glenoid fossa of scapula	Radial tuberosity	Flexes elbow joint; supinates forearm and hand at radioulnar joint
Brachialis	Anterior body of humerus	Coronoid process of ulna	Flexes elbow joint
Brachioradialis	Lateral supracondylar ridge of humerus	Proximal to styloid process of radius	Flexes elbow joint
Triceps brachii	Tuberosity below glenoid fossa: lateral and medial surfaces of humerus	Olecranon of ulna	Extends elbow joint
Anconeus	Lateral epicondyle of humerus	Olecranon of ulna	Extends elbow joint

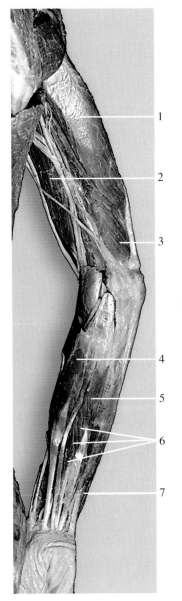

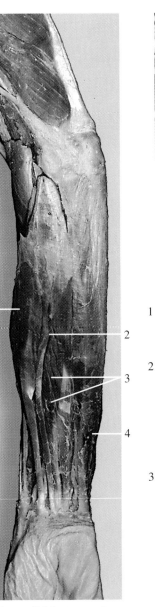

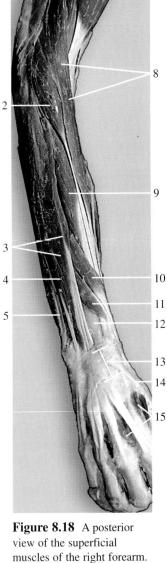

Figure 8.15 The medial brachium and superficial flexors of the right antebrachium.
1. Triceps brachii m. (long head)
2. Biceps brachii m. (short head)
3. Triceps brachii m. (medial head)
4. Flexor carpi radialis m.
5. Palmaris longus m.
6. Superficial digital flexor m.
7. Flexor carpi ulnaris m.

Figure 8.16 An anterior view of the superficial muscles of the right forearm.
1. Flexor carpi radialis m.
2. Palmaris longus m.
3. Superficial digital flexor m.
4. Flexor carpi ulnaris m.

Figure 8.17 An anterior view of the deep muscles of the right forearm.
1. Pronator teres m.
2. Flexor pollicis longus m.
3. Pronator quadratus m.
4. Median nerve
5. Deep digital flexor m.

Figure 8.18 A posterior view of the superficial muscles of the right forearm.
1. Triceps brachii m. (medial head)
2. Extensor carpi radialis longus m.
3. Extensor digitorum m.
4. Extensor carpi minimi m.
5. Extensor carpi carpi ulnaris m.
6. Brachialis m.
7. Biceps brachii m. (long head)
8. Brachioradialis m.
9. Extensor carpi radialis brevis m.
10. Abductor pollicis longus m.
11. Extensor pollicis brevis m.
12. Radius
13. Extensor retinaculum
14. Tendon of extensor pollicis longus m.
15. Dorsal interosseous mm.

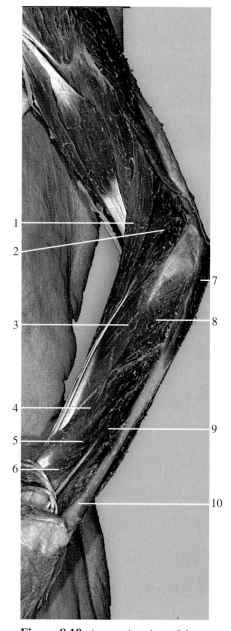

Figure 8.19 A posterior view of deep muscles of the left forearm.

1. Brachioradialis m.
2. Extensor carpi radialis longus m.
3. Extensor carpi radialis brevis m.
4. Abductor pollicis longus m.
5. Extensor pollicis brevis m.
6. Extensor pollicis longus m.
7. Anconeus m.
8. Supinator m.
9. Extensor indicis m.
10. Ulna

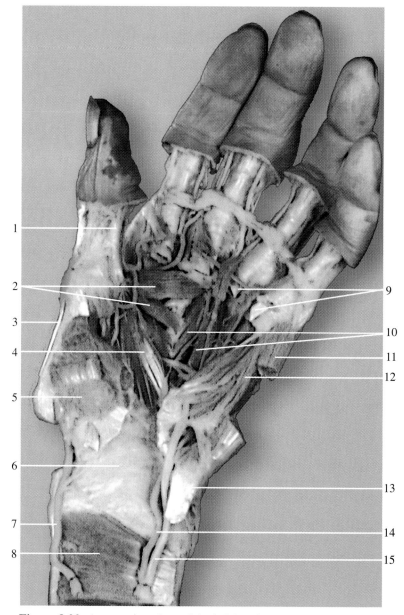

Figure 8.20 An anterior view of the left hand

1. Tendon of flexor pollicis longus muscle
2. Adductor pollicis muscles
3. Tendon of exterior pollicis brevic muscle
4. Flexor pollicis brevis muscle
5. Opponens pollicis muscle
6. Flexor synovial sheath
7. Radial artery
8. Pronator quadratus muscle
9. Tendons of deep digital flexor muscle (cut)
10. Lumbricals
11. Abductor for digiti minimi and flexor digiti minimi muscles
12. Opponens digiti minimi muscle
13. Tendon of flexor carpi ulnaris muscle
14. Ulnar artery
15. Ulnar nerve

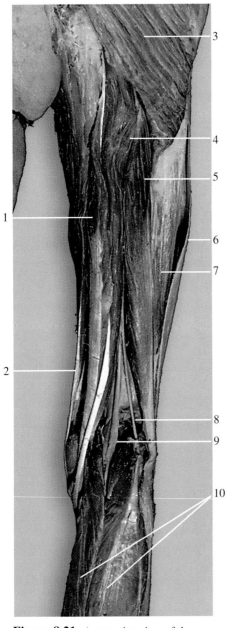

Figure 8.21 A posterior view of the right thigh.
1. Semimembranosus m.
2. Gracilis m.
3. Gluteus maximus m.
4. Semitendinosus m.
5. Biceps femoris (long head)
6. Iliotibial tract
7. Vastus lateralis m.
8. Common peroneal nerve
9. Tibial nerve
10. Gastrocnemius m. (lateral and medial heads)

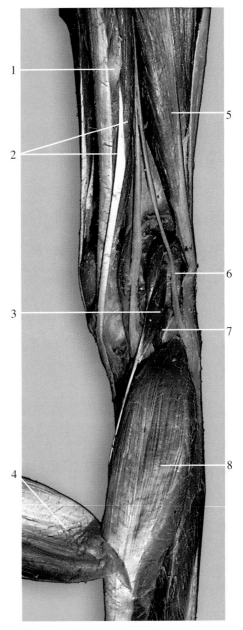

Figure 8.22 The right popliteal fossa and the surrounding structures.
1. Semimembranosus m.
2. Semitendinosus m
3. Plantaris m.
4. Gastrocnemius m. (reflected)
5. Biceps femoris m.
6. Common peroneal nerve
7. Popliteus m.
8. Soleus m.

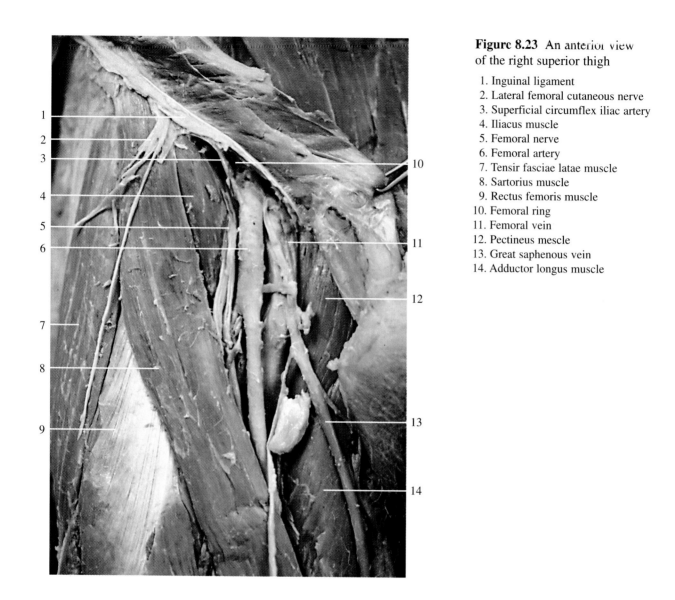

Figure 8.23 An anterior view of the right superior thigh

1. Inguinal ligament
2. Lateral femoral cutaneous nerve
3. Superficial circumflex iliac artery
4. Iliacus muscle
5. Femoral nerve
6. Femoral artery
7. Tensir fasciae latae muscle
8. Sartorius muscle
9. Rectus femoris muscle
10. Femoral ring
11. Femoral vein
12. Pectineus mescle
13. Great saphenous vein
14. Adductor longus muscle

Table 8.8 Medial Muscles that Move the Thigh at the Hip Joint

Adductor muscle	Origin(s)	Insertion(s)	Action
Gracilis	Inferior edge of symphysis pubis	Proximomedial surface of tibia	Adducts thigh at hip joint; flexes and rotates leg at knee joint
Pectineus	Pectineal line of pubis	Distal to lesser trochanter of femur	Adducts and flexes thigh at hip joint
Adductor longus	Pubis—below pubic crest	Linea aspera of femur	Adducts, flexes, and laterally rotates thigh at hip joint
Adductor brevis	Inferior ramus of pubis	Linea aspera of femur	Adducts, flexes, and laterally rotates thigh at hip joint
Adductor magnus	Inferior ramus of ischium and inferior ramus of pubis	Linea aspera and medial epicondyle of femur	Adducts, flexes, and laterally rotates thigh at hip joint

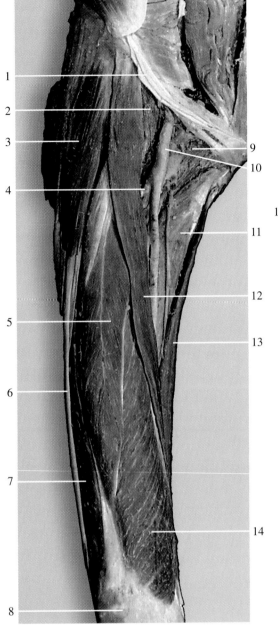

Figure 8.24 An anterior view of the right thigh.

1. Inguinal ligament
2. Iliopsoas m.
3. Tensor faciae latae m.
4. Deep femoral artery
5. Rectus femoris m.
6. Iliotibial tract
7. Vastus lateralis m.
8. Patella
9. Pectineus m.
10. Femoral artery
11. Adductor longus m.
12. Sartorius m.
13. Gracilis m.
14. Vastus medialis m.

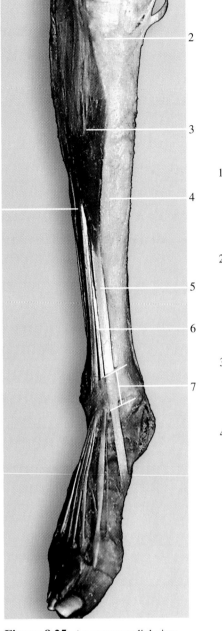

Figure 8.25 An anteromedial view of the muscles of the right leg.

1. Extensor digitorum longus m.
2. Tuberosity of the tibia
3. Tibialis anterior m.
4. Tibia
5. Tendon of tibialis anterior m.
6. Tendon of extensor hallucis longus m.
7. Superior extensor retinaculum

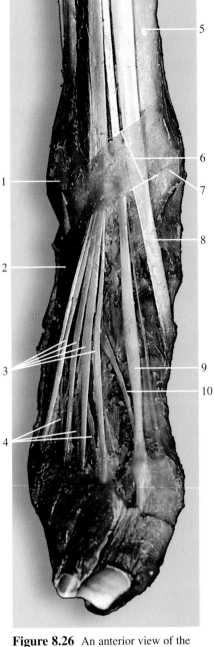

Figure 8.26 An anterior view of the dorsum of the right foot.

1. Lateral malleolus
2. Extensor digitorum brevis m.
3. Tendons of extensor digitorum longus m.
4. Tendons of extensor digitorum brevis m.
5. Tibia
6. Superior extensor retinaculum
7. Medial malleolus
8. Tendon of tibialis anterior m.
9. Tendon of extensor hallucis longus m.
10. Tendon of extensor hallucis brevis m.

Table 8.9 Muscles that Act on the Wrist, Hand, and Fingers

Antebrachial muscle	Origin(s)	Insertion(s)	Action
Supinator	Lateral epicondyle of humerus and crest of ulna	Lateral surface of radius	Supinates forearm and hand
Pronator teres	Medial epicondyle of humerus	Lateral surface of radius	Pronates forearm and hand
Pronator quadratus	Distal fourth of ulna	Distal fourth of radius	Pronates forearm and hand
Flexor carpi radialis	Medial epicondyle of humerus	Base of second and third metacarpal bones	Flexes and abducts hand at wrist
Palmaris longus	Medial epicondyle of humerus	Palmar aponeurosis	Flexes wrist
Flexor carpi ulnaris	Medial epicondyle of humerus and olecranon of ulna	Carpal and metacarpal bones	Flexes and adducts wrist
Superficial digital flexor	Medial epicondyle of humerus and coronoid process of ulna	Middle phalanges of digits II–V	Flexes wrist and digits
Deep digital flexor	Proximal two-thirds of ulna and interosseous ligament	Distal phalanges of digits II–V	Flexes wrist and digits
Flexor pollicis longus	Body of radius and coronoid process of ulna	Distal phalanx of thumb	Flexes joints of thumb
Extensor carpi radialis longus	Lateral epicondyle of humerus	Second metacarpal bone	Extends and abducts wrist
Extensor carpi radialis brevis	Lateral epicondyle of humerus	Third metacarpal bone	Extends and abducts wrist
Extensor digitorum communis	Lateral epicondyle of humerus	Posterior surfaces of digits II–V	Extends wrist and phalanges
Extensor digiti minimi	Lateral epicondyle of humerus	Extensor aponeurosis of fifth digit	Extends joints of fifth digit and wrist
Extensor carpi ulnaris	Lateral epicondyle of humerus and olecranon of ulna	Base of fifth metacarpal bone	Extends and adducts wrist
Extensor pollicis longus	Lateromedial body of ulna	Lateromedial body of ulna	Extends joints of thumb; abducts joints of hand
Extensor pollicis brevis	Distal body of radius and interosseous ligament	Base of first phalanx of thumb	Extends joints of thumb; abducts joints of hand
Abductor pollicis longus	Distal radius and ulna and interosseous ligament	Base of first metacarpal bone	Abducts joints of thumb and joints of hand

Table 8.10 Muscles that Act on the Thigh at the Hip Joint

Pelvic muscle	*Origin(s)*	*Insertion(s)*	*Action*
Iliacus	Iliac fossa	Lesser trochanter of femur, along with psoas major	Flexes and rotates thigh laterally at the hip joint; flexes joints of vertebral column
Psoas major	Transverse process of lumbar vertebrae	Lesser trochanter of femur, along with iliacus	Flexes and rotates thigh laterally at the hip joint; flexes joints of vertebral column
Gluteus maximus	Iliac crest, sacrum, coccyx, aponeurosis of lumbar region	Gluteal tuberosity and iliotibial tract	Extends and rotates thigh laterally at the hip joint
Gluteus medius	Lateral surface of ilium	Greater trochanter of femur	Abducts and rotates thigh medially at the hip joint
Gluteus minimus	Lateral surface of lower half of ilium	Greater trochanter of femur	Abducts and rotates thigh medially at the hip joint
Tensor fasciae latae	Anterior border of ilium and iliac crest	Iliotibial tract	Abducts thigh at the hip joint

Table 8.11 Muscles of the Thigh that Act on the Leg

Thigh muscle	*Origin(s)*	*Insertion(s)*	*Action*
Sartorius	Anterior superior iliac spine	Medial surface of tibia	Flexes leg and thigh; abducts and rotates thigh laterally; rotates leg medially at hip joint
Quadriceps femoris		Patella by common tendon, which continues as patellar ligament to tibial tuberosity	Extends leg at knee joint
Rectus femoris	Anterior inferior iliac spine		
Vastus lateralis	Greater trochanter and linea aspera of femur		
Vastus medialis	Medial surface and linea aspera of femur		
Vastus intermedius	Anterior and lateral surfaces of femur		
Biceps femoris	Long head—ischial tuberosity; short head—linea aspera of femur	Head of fibula and proximolateral part of tibia	Flexes leg at knee joint; extends and laterally rotates thigh at hip joint
Semitendinosus	Ischial tuberosity	Proximomedial surface of tibia	Flexes leg at knee joint; extends and medially rotates thigh at hip joint
Semimembranosus	Ischial tuberosity	Proximomedial part of tibia	Flexes leg at knee joint; extends and medially rotates thigh at hip joint

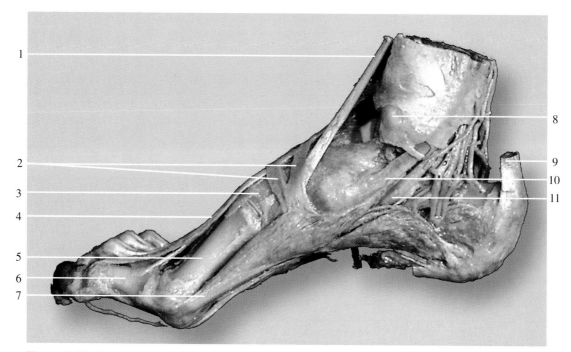

Figure 8.27 A medial view of the right foot
1. Tendon of tibialis anterior muscle
2. Extensor retinaculum
3. Medial cuneiform
4. Tendon of extensor digitorum longus muscle
5. First metatarsal bone
6. Proximal phalanx of hallux

7. Abductor hallucis muscle
8. Medial malleolus of tibia
9. Tendo calcaneus
10. Tendon of tibialis posterior muscle
11. Tendon of flexor digitorum longus muscle

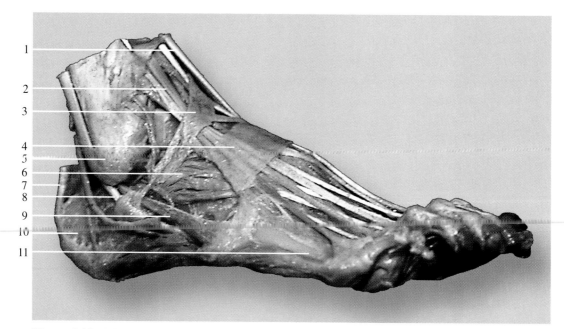

Figure 8.28 A lateral view of the right foot
1. Tendon of tibialis anterior muscle
2. Tendon of extensor digitorum longus muscle
3. Superior extensor retinaculum
4. Inferior extensor retinaculum
5. Lateral malleolus of fibula
6. Extensor digitorum brevis muscle

7. Tendo calcaneus
8. Tendon of peroneus longus muscle
9. Tendon of peroneus brevis muscle
10. Calcaneus
11. Fifth metatarsal bone

Table 8.12 Muscles of the Leg That Move the Ankle, Foot, and Toes

Leg muscle	*Origin(s)*	*Insertion(s)*	*Action*
Tibialis anterior	Lateral condyle and body of tibia	First metatarsal bone and first cuneiform bone	Dorsiflexes ankle; inverts foot and ankle
Extensor digitorum longus	Lateral condyle of tibia and anterior surface of fibula	Extensor expansions of digits II–V	Extends digits II–V; dorsiflexes foot at ankle
Extensor hallucis longus	Anterior surface of fibula and interosseous ligament	Distal phalanx of digit I	Extends joints of big toe; assists dorsiflexion of foot at ankle
Peroneus tertius	Anterior surface of fibula and interosseous ligament	Dorsal surface of fifth metatarsal bone	Dorsiflexes and everts foot at ankle
Peroneus longus	Lateral condyle of tibia and head and shaft of fibula	First cuneiform and metatarsal bone I	Plantar flexes and everts foot at ankle
Peroneus brevis	Lower aspect of fibula	Metatarsal bone V	Plantar flexes and everts foot at ankle
Gastrocnemius	Lateral and medial condyle of femur	Posterior surface of calcaneous	Plantar flexes foot at ankle; flexes knee joint
Soleus	Posterior aspect of fibula and tibia	Calcaneous	Plantar flexes foot at ankle
Plantaris	Supracondylar ridge of femur	Calcaneous	Plantar flexes foot at ankle
Popliteus	Lateral condyle of femur	Upper posterior aspect of tibia	Flexes and medially rotates leg at knee joint
Flexor hallucis longus	Posterior aspect of fibula	Distal phalanx of big toe	Flexes joint of distal phalanx of big toe
Flexor digitorum longus	Posterior surface of tibia	Distal phalanges of digits II–V	Flexes joints of distal phalanges of digits II–V
Tibialis posterior	Tibia and fibula and interosseous ligament	Navicular, cuneiform, cuboid, and metatarsal bones II–IV	Plantar flexes and inverts foot at ankle; supports arches of foot

Nervous System

9

The nervous system is anatomically divided into the *central nervous system (CNS)*, which includes the *brain* and *spinal cord*, and the *peripheral nervous system (PNS)*, which includes the *cranial nerves*, arising from the brain, and the *spinal nerves*, arising from the spinal cord (fig. 9.1). The *autonomic nervous system (ANS)* is a functionally distinct division of the nervous system devoted to regulation of involuntary activities in the body. The ANS is made up of specific portions of the CNS and PNS.

The brain and spinal cord are the centers for integration and coordination of information. Conveyed as *nerve impulses*, information to and from the brain and spinal cord travels through nerves. *Nerves* are similar to electrical conducting wires. Nerve impulses are sent from the brain in the form of electrical signals along *motor nerves* to the receiving organs, which then translate the signal into some specific function. For example, the motor impulses conducted from the brain to the muscles of the forearm that serve the hand cause the fingers to move as the muscles are contracted. *Sensory nerves* conduct nerve impulses in the opposite direction—from the receptor site to the CNS. For example, a pinprick on the skin produces a sensory impulse along a sensory nerve that the brain interprets as a painful sensation.

Neurons and *neuroglia* are the two cell types that make up nervous tissue. Neurons are specialized to respond to physical and chemical stimuli, conduct impulses, and release specific chemical regulators, called *neurotransmitters*. Although neurons vary considerably in size and shape, they have three principal components: a *cell body, dendrites,* and an *axon* (fig. 9.3). In a typical neuron connection, the axon of one neuron *synapses* (joins) on the cell body or dendrites of a neighboring neuron. Axons vary in length from a few millimeters in the CNS to over a meter in the PNS. Long axons are generally *myelinated* with *neurolemmocytes (Schwann cells)* in the PNS, and many of the short axons are myelinated with *oligodendrocytes* in the CNS. *Neurofibril nodes (nodes of Ranvier)* are segments in the *myelin sheath.* The end of the axon at the synapse is called the *axon terminal.*

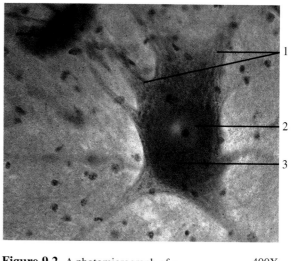

Figure 9.2 A photomicrograph of a neuron.   400X
1. Cytoplasmic extensions
2. Nucleus
3. Cell body of neuron

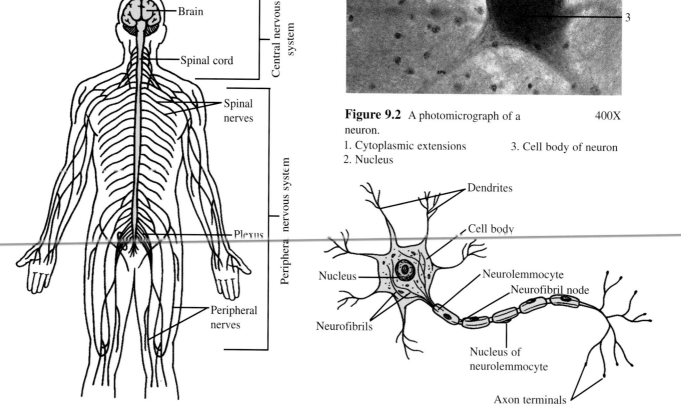

Figure 9.1 Divisions of the nervous system.

Figure 9.3 Structure of a myelinated neuron.

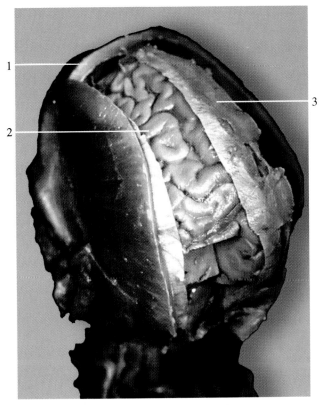

Figure 9.4 A sectioned cranium exposing the meninges and cerebrum.
1. Skull   3. Dura mater
2. Cerebrum

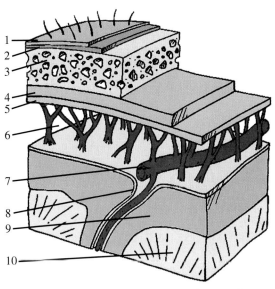

Figure 9.5 The relationship of the meninges to the skull and the cerebrum.
1. Skin of scalp   6. Subarachnoid space
2. Galea aponeurotica   7. Blood vessel
3. Bone of cranium   8. Pia mater
4. Dura mater   9. Cerebral cortex
5. Arachnoid   10. White matter of brain

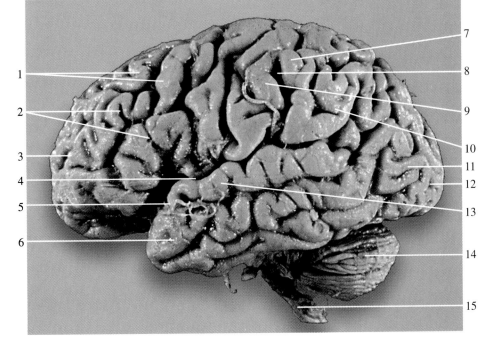

Figure 9.6 A lateral view of the brain.
1. Gyri
2. Sulci
3. Frontal lobe of cerebrum
4. Lateral sulcus
5. Olfactory cerebral cortex
6. Temporal lobe of cerebrum
7. Primary sensory cerebral cortex
8. Central sulcus
9. Primary motor cerebral cortex
10. Parietal lobe of cerebrum
11. Occipital lobe of cerebrum
12. Visual cerebral cortex
13. Auditory cerebral cortex
14. Cerebellum
15. Medulla oblongata

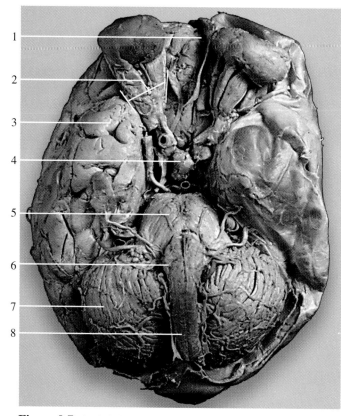

Figure 9.7 An inferior view of the brain with the eyes and part of the meninges still intact.

1. Longitudinal cerebral fissure
2. Muscles of the eye
3. Temporal lobe of cerebrum
4. Pituitary gland
5. Pons
6. Medulla oblongata
7. Cerebellum
8. Spinal cord

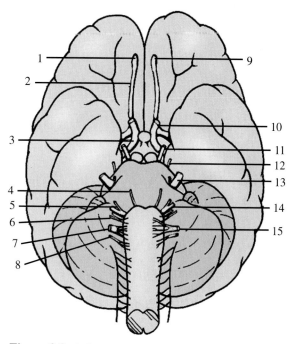

Figure 9.8 A diagram of the inferior of the brain showing the cranial nerves.

1. Olfactory bulb
2. Olfactory tract
3. Optic tract
4. Abducens nerve
5. Facial nerve
6. Glossopharyngeal nerve
7. Vagus nerve
8. Accessory nerve
9. Olfactory nerve
10. Optic nerve
11. Oculomotor nerve
12. Trochlear nerve
13. Trigeminal nerve
14. Vestibulocochlear nerve
15. Hypoglossal nerve

Figure 9.9 Cranial nerves and blood supply to the brain.

1. Internal carotid artery
2. Cerebral arterial circle (circle of Willis)
3. Trigeminal nerve
4. Vestibulocochlear nerve
5. Abducens nerves
6. Hypoglossal nerve
7. Olfactory tract
8. Optic nerve
9. Oculomotor nerve
10. Trochlear nerve
11. Trigeminal nerve
12. Facial nerve
13. Glossopharyngeal nerve
14. Vagus nerve
15. Vertebral artery

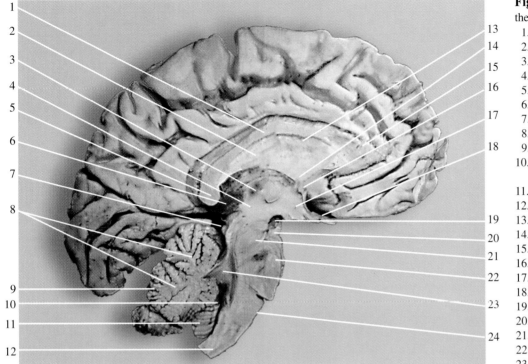

Figure 9.10 A sagittal view of the brain.

1. Truncus of corpus callosum
2. Crus of fornix
3. Third ventricle
4. Posterior commissure
5. Splenium of corpus callosum
6. Pineal body
7. Inferior colliculus
8. Arbor vitae of cerebellum
9. Vermis of cerebellum
10. Choroid plexus of fourth ventricle
11. Tonsilla of cerebellum
12. Medulla oblongata
13. Septum pellucidum (cut)
14. Intraventricular foramen
15. Genu of corpus callosum
16. Anterior commissure
17. Hypothalmus
18. Optic chiasma
19. Oculomotor nerve
20. Cerebral peduncle
21. Midbrain
22. Pons
23. Fourth ventricle
24. Pyramid

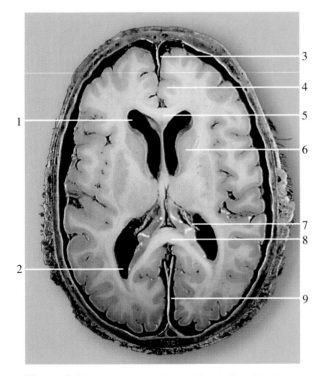

Figure 9.11 Transaxial section of the skull and brain.

1. First ventricle
2. Second ventricle
3. Falx cerebri (septum of dura mater)
4. Cingulate gyrus
5. Genu of corpus callosum
6. Caudate nucleus
7. Choroid plexus
8. Splenium of corpus callosum
9. Falx cerebri (septum of dura mater)

Figure 9.12 Coronal MRI brain scan.

1. Cerebrospinal fluid
2. Longitudinal cerebral fissure
3. Third ventricle
4. Cerebellum
5. Skull
6. Dura mater
7. Cerebral cortex
8. Cerebral medulla
9. Lateral ventricle
10. Fourth ventricle

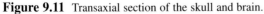

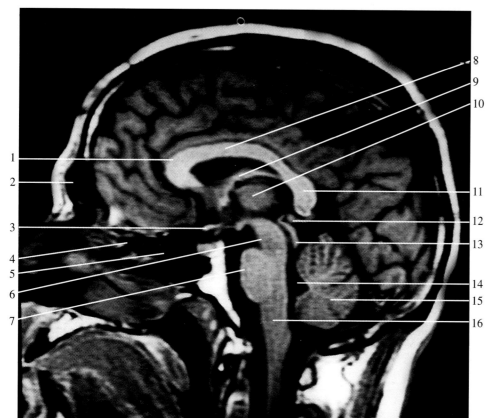

Figure 9.13 MRI sagittal section through the skull.

1. Genu of corpus callosum
2. Frontal sinus
3. Pituitary gland
4. Ethmoidal sinus
5. Sphenoidal sinus
6. Tegmentum (midbrain)
7. Pons
8. Truncus of corpus callosum
9. Fornix
10. Thalamus
11. Splenium of corpus callosum
12. Pineal gland
13. Superior and inferior colliculi
14. Fourth ventricle
15. Cerebellum
16. Medulla oblongata

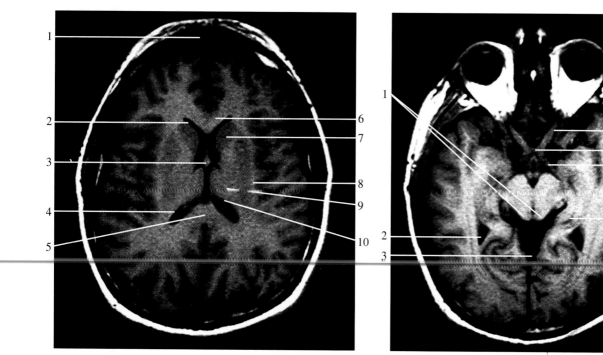

Figure 9.14 MRI transaxial section through the brain.

1. Frontal sinus
2. Frontal horn of lateral ventricle
3. Body of fornix
4. Posterior horn of lateral ventricle
5. Splenium of corpus callosum
6. Genu of corpus callosum
7. Head of caudate nucleus
8. External capsule
9. Thalamus
10. Choroid plexus

Figure 9.15 MRI transaxial section showing visual pathways.

1. Superior colliculi
2. Lateral ventricle
3. Third ventricle
4. Optic nerve
5. Optic chiasma
6. Optic tract
7. Lateral geniculate body
8. Calcarine tracts (optic radiation)

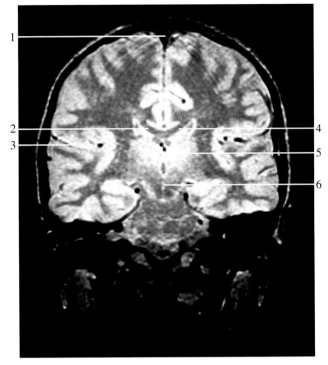

Figure 9.16 A MRI coronal section through the thalamus.
1. Superior sagittal sinus   4. Corpus callosum
2. Lateral ventricle   5. Thalamus
3. Lateral fissure   6. Third ventricle

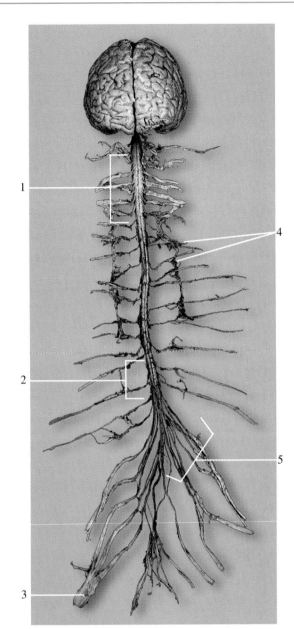

Figure 9.18 Anterior surface of the brain and spinal cord with meninges removed.
1. Cervical enlargement
2. Lumbar enlargement
3. Sciatic nerve
4. Sympathetic ganglia
5. Cauda equina

Figure 9.17 A posterior view of the lower spinal cord.
1. Spinal cord
2. Dura mater (cut)
3. Posterior (dorsal) root of spinal nerve
4. Cauda equina
5. Filum terminale

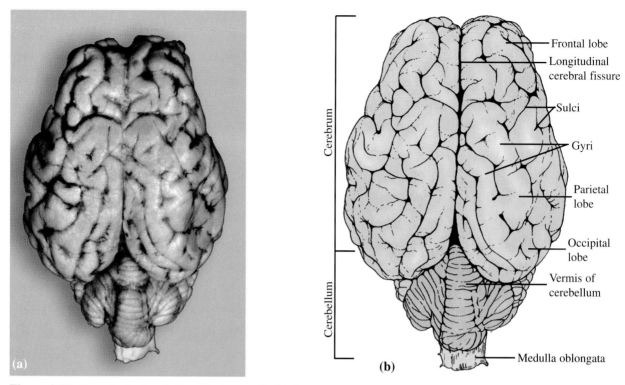

Figure 9.19 Sheep brain, dorsal view. (a) photograph; (b) diagram.

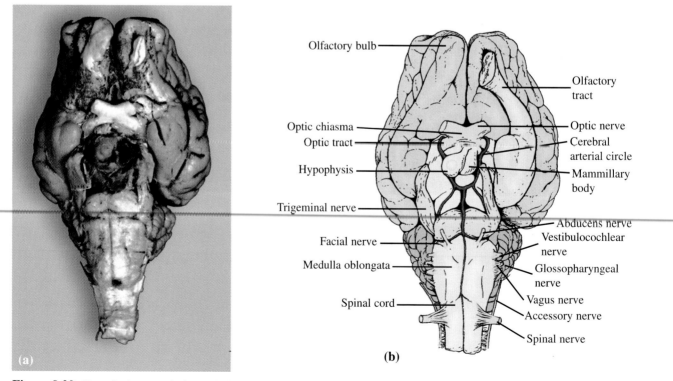

Figure 9.20 Sheep brain, ventral view. (a) photograph; (b) diagram.

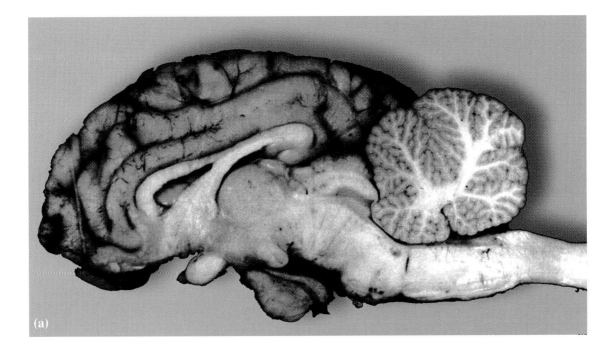

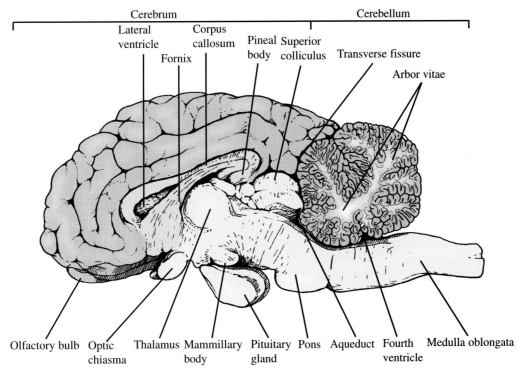

Figure 9.21 Sheep brain, sagittal view. (a) photograph; (b) diagram.

Endocrine System

10

The endocrine system works closely with the nervous system to regulate and integrate body processes and maintain homeostasis. The nervous system regulates body activities through the action of electrochemical impulses that are transmitted by means of neurons, resulting in rapid, but usually brief responses. By contrast, the endocrine system is composed of glands (fig. 10.1) scattered throughout the body that release chemical substances called *hormones* into the bloodstream. These hormones dissipate in the blood and travel throughout the entire body to act on *target tissues*, where they have a slow but relatively long-lasting effect. Neurological responses are measured in milliseconds, but hormonal action requires seconds or days to elicit a response. Some hormones may have an effect that lasts for minutes and others, for weeks or months.

The endocrine system and nervous system are closely coordinated in autonomically controlling the functions of the body. The *pituitary gland*, located in the brain, regulates the activity of most other endocrine glands. Located immediately between the pituitary and the rest of the brain is the *hypothalamus* The hypothalamus serves as an intermediate between the nervous centers of the brain and the pituitary gland, correlating the activity of the two systems. Furthermore, certain hormones may stimulate or inhibit the activities of the nervous system.

Other organs of the endocrine system include the *thyroid gland* and *parathyroid glands*, located in the neck. The *adrenal glands* and *pancreas* are located in the abdominal region. The *ovaries* of the female are located in the pelvic cavity, whereas the *testes* of the male are located in the scrotum. Even the *placenta* serves as an endocrine organ for the developing fetus and has some hormonal influence upon the pregnant woman.

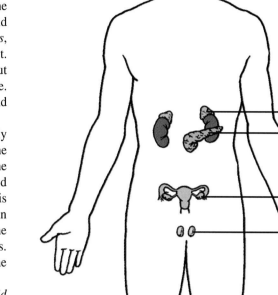

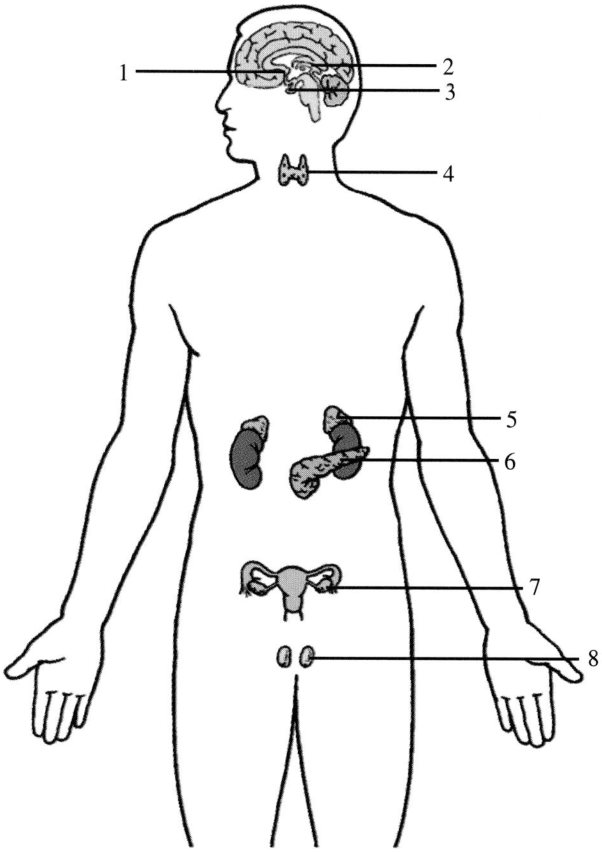

Figure 10.1 Principal endocrine glands.

1. Hypothalamus
2. Pineal body
3. Pituitary gland
4. Thyroid and parathyroid glands
5. Adrenal (suprarenal) gland
6. Pancreas
7. Ovary
8. Testis

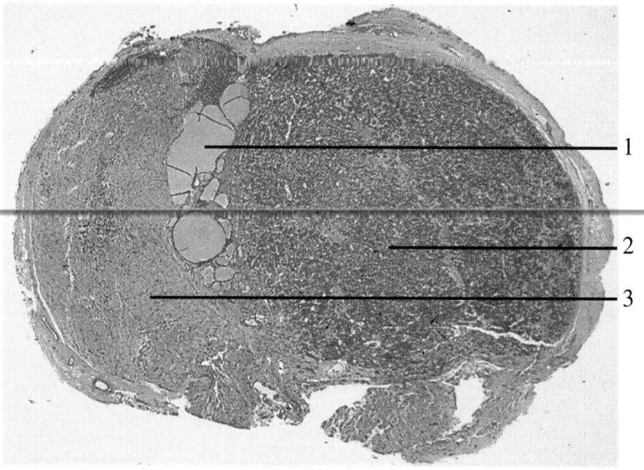

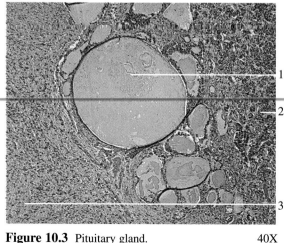

Figure 10.2 Pituitary gland.   7X
1. Pars intermedia (adenohypophysis)
2. Pars distalis (adenohypophysis)
3. Pars nervosa (neurohypophysis)

Figure 10.3 Pituitary gland.   40X
1. Pars intermedia
2. Pars distalis
3. Pars nervosa

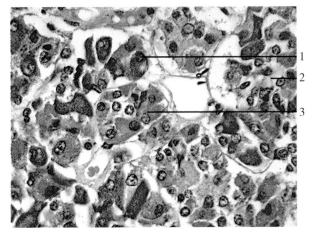

Figure 10.4 Pars distalis of the pituitary gland. 400X
1. Basophil 3. Acidophil
2. Chromophobe

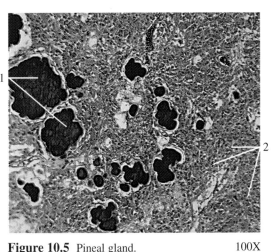

Figure 10.5 Pineal gland. 100X
1. Brain sand
2. Pinealocytes

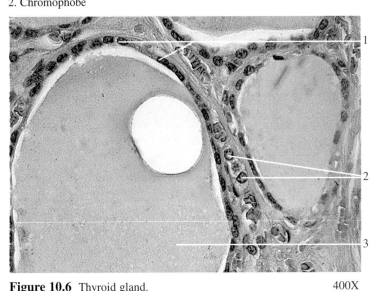

Figure 10.6 Thyroid gland. 400X
1. Follicle cells 3. Colloid within follicle
2. C cells

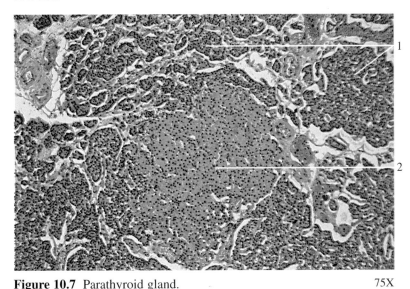

Figure 10.7 Parathyroid gland. 75X
1. Chief cells 2. Cluster of oxyphil cells

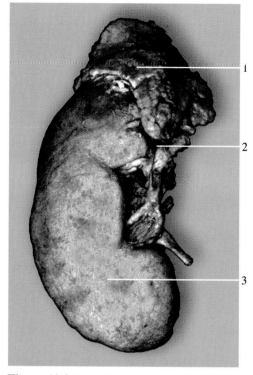

Figure 10.8 The adrenal (suprarenal) gland.
1. Adrenal gland
2. Inferior suprarenal artery
3. Kidney

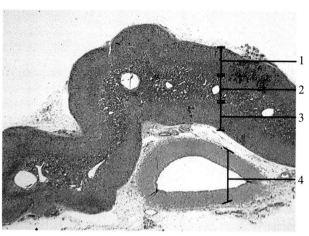

Figure 10.9 Adrenal gland.   7X
1. Adrenal cortex
2. Adrenal medulla
3. Adrenal cortex
4. Blood vessel

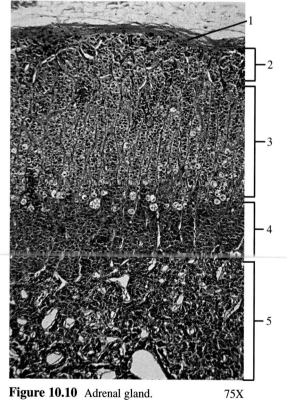

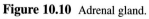

Figure 10.10 Adrenal gland.   75X
1. Capsule
2. Zona glomerulosa (adrenal cortex)
3. Zona fasiculata (adrenal cortex)
4. Zona reticularis (adrenal cortex)
5. Adrenal medulla

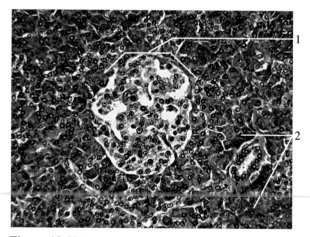

Figure 10.11 Pancreatic islet (islet of Langerhans).   75X
1. Pancreatic islet (endocrine pancreas)
2. Acini (exocrine pancreas)

Sensory Organs

11

The nervous and endocrine systems convey information from the brain to all parts of the body to enable a person to interact with both the external and internal environments and to maintain homeostasis. The sense organs, in contrast, convey information from the outside world (and inside world of the body) back to the brain. This includes a wide range of information such as temperature, brightness, sound, flavor, and balance.

The sense organs are actually extensions of the nervous system that allow us to autonomically respond or conscientiously perceive our internal and external environments. A stimulus excites a sense organ which then transduces the stimulus to an electrical (nerve) impulse. Sensory nerves transmit the impulse (sensation) to the brain to be perceived and acted upon. Ultimately, it is the brain which actually feels, sees, hears, tastes, and smells.

The *eyes* are the organs of visual sense. The eyes refract (bend) and focus the incoming light waves onto the sensitive *photoreceptors (rods and cones)* at the back of each eye. Nerve impulses from the stimulated photoreceptors are conveyed along visual pathways to the occipital lobes of the cerebrum, where visual sensations are perceived.

The eyeball consists of the fibrous tunic, which is divided into the *sclera* and *cornea*; the vascular tunic, which consists of the *choroid*, the *ciliary body*, and the *iris*; and the internal tunic, or *retina*, which consists of an outer pigmented layer and an inner nervous layer. The eye contains an anterior cavity between the lens and the retina. The anterior cavity is subdivided into an anterior chamber in front of the iris and a posterior chamber behind the iris. *Aqeous humor* fills both of these chambers. The posterior cavity (also called the vitreous chamber) contains vitreous humor.

The *ear* is the organ of hearing and equilibrium (balance). It contains receptors that respond to movements of the head and receptors that convert sound waves into nerve impulses. Impulses from both receptor types are transmitted through the vestibulocochlear (VIII) cranial nerve to the brain for interpretation.

The ear consists of the three principal regions: the *outer ear*, the *middle ear*, and the *inner ear*. The outer ear consists of the *auricle* and the *external auditory canal*. The middle ear contains the auditory ossicles (*malleus*, *incus*, and *stapes*). The inner ear contains the *spiral organ* (organ of Corti) in the *cochlea* for hearing, and the *semicircular canals* for equilibrium.

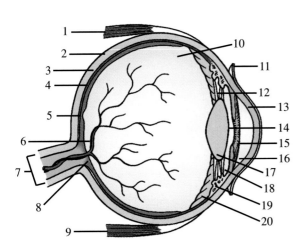

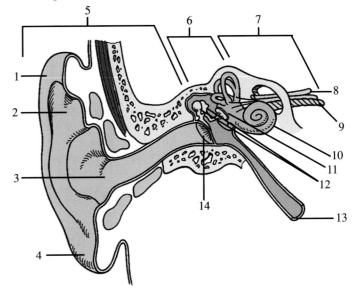

Figure 11.1 Structure of the eye.

1. Superior rectus m.	11. Conjuctiva
2. Sclera	12. Suspensory ligament
3. Choroid	13. Cornea
4. Retina	14. Pupil
5. Fovea centralis	15. Iris
6. Central vessels	16. Anterior chamber
7. Optic nerve	17. Lens
8. Optic disc	18. Posterior chamber
9. Inferior rectus m.	19. Ciliary body
10. Vitreous chamber	20. Ora serrata

Figure 11.2 Structure of the ear.

1. Helix	8. Facial nerve
2. Auricle	9. Vestibulocochlear nerve
3. External auditory canal	10. Cochlea
4. Earlobe	11. Vestibular (oval) window
5. Outer ear	12. Auditory ossicles
6. Middle ear	13. Auditory tube
7. Inner ear	14. Tympanic membrane

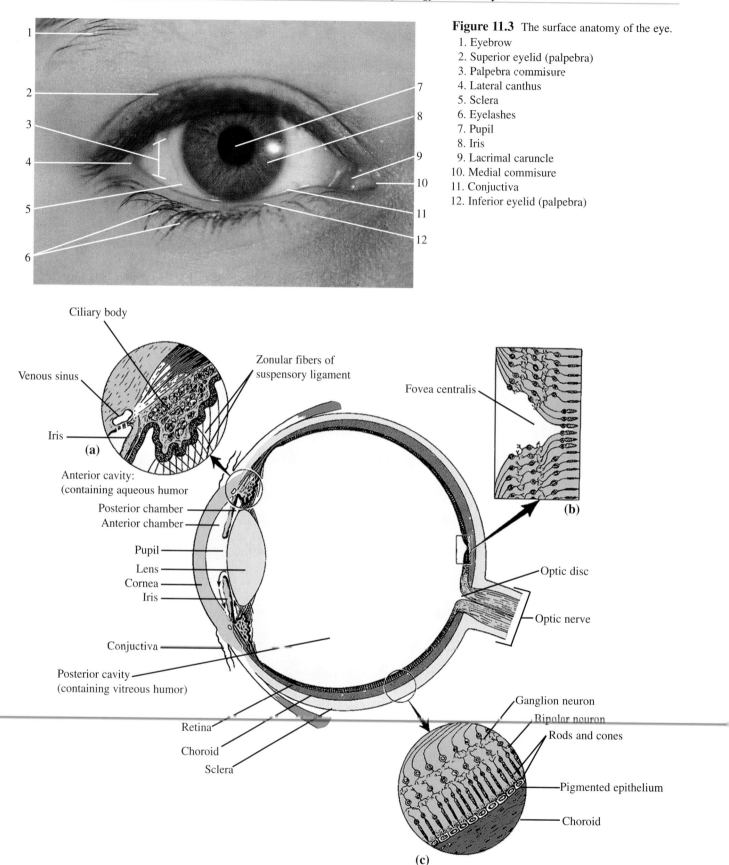

Figure 11.3 The surface anatomy of the eye.
1. Eyebrow
2. Superior eyelid (palpebra)
3. Palpebra commisure
4. Lateral canthus
5. Sclera
6. Eyelashes
7. Pupil
8. Iris
9. Lacrimal caruncle
10. Medial commisure
11. Conjuctiva
12. Inferior eyelid (palpebra)

Figure 11.4 Structure of the eye. (a) The ciliary body, (b) fovea centralis, and (c) retina.

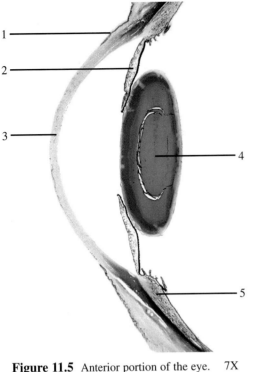

Figure 11.5 Anterior portion of the eye.   7X
1. Conjunctivia   4. Lens
2. Iris   5. Ciliary body
3. Cornea

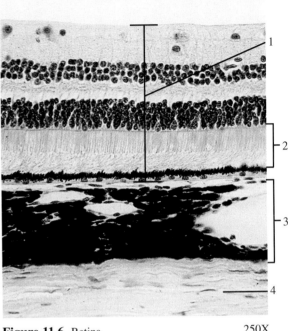

Figure 11.6 Retina.   250X
1. Retina   3. Choroid
2. Rods and cones   4. Sclera

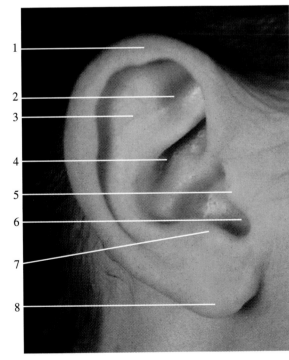

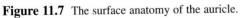

Figure 11.7 The surface anatomy of the auricle.
1. Helix   5. Tragus
2. Triangular fossa   6. External auditory canal
3. Antihelix   7. Antitragus
4. Concha   8. Earlobe

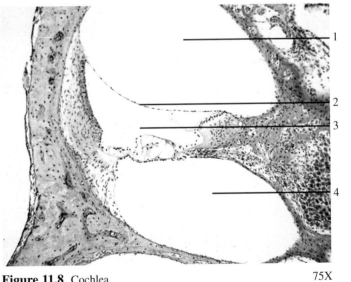

Figure 11.8 Cochlea.   75X
1. Scala vestibuli   3. Cochlear duct
2. Vestibular membrane   4. Scala tympani

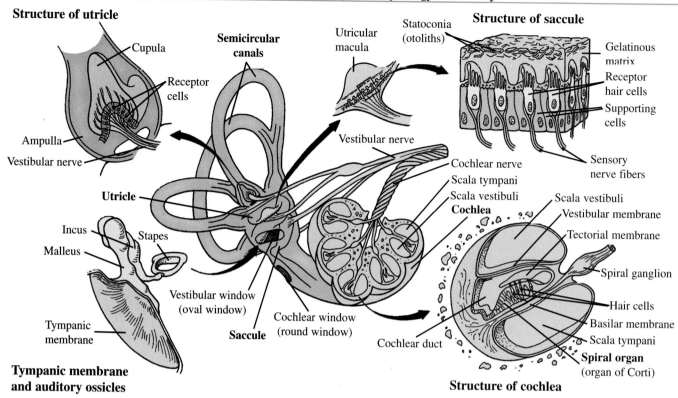

Figure 11.9 Structures of the middle ear and inner ear. The tympanic membrane and auditory ossicles (malleus, incus, stapes) are structures of the middle ear. The vestibular organs (utricle, saccule, semicircular canals) and cochlea (containing the spiral organ) are structures of the inner ear.

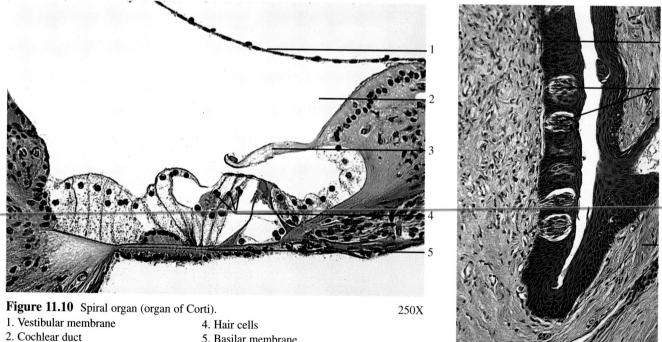

Figure 11.10 Spiral organ (organ of Corti).   250X

1. Vestibular membrane
2. Cochlear duct
3. Tectorial membrane
4. Hair cells
5. Basilar membrane

Figure 11.11 Taste bud.   200X
1. Epithelium of papilla
2. Taste buds
3. Tongue muscle

Circulatory System

12

The circulatory system consists of the *blood*, *heart*, and *vessels*, each of which is essential to the life of a complex multicellular organism. Blood, a specialized connective tissue, consists of *formed elements* (erythrocytes, leukocytes, and thrombocytes) that are suspended and carried in the *plasma*. These formed elements function in transport, immunity, and blood-clotting mechanisms.

The heart is enclosed in a *pericardial sac* within the thoracic cavity. The wall of the heart consists of the *epicardium*, *myocardium*, and *endocardium*. The *right atrium* of the heart receives blood from the superior vena cava and inferior vena cava, and the *right ventricle* pumps blood into the *pulmonary trunk* to the *pulmonary arteries*. The *left atrium* receives blood from the *pulmonary veins* and pumps blood into the left ventricle. The left ventricle pumps blood into the *aorta*.

There are four heart valves that prohibit the backflow of blood: 1) The *right tricuspid valve (right atrioventricular valve)* is located between the right atrium and the right ventricle; 2) the *pulmonary semilunar valve (pulmonary valve)* is located between the right ventricle and the pulmonary trunk; 3) the *left bicuspid valve (left atrioventricular, or mitral, valve)* is located between the left atrium and the left ventricle; and, 4) the *aortic semilunar valve (aortic valve)* is located between the left ventricle and the ascending aorta.

The *systemic arteries* arise from the aorta or branches of the aorta and transport blood away from the heart to smaller vessels called *arterioles*. From arterioles, the blood enters *capillaries* where diffusion with the surrounding cells may occur. Capillaries converge forming *venules*, which in turn converge forming larger vessels called *veins*. Veins are vessels that transport blood toward the heart.

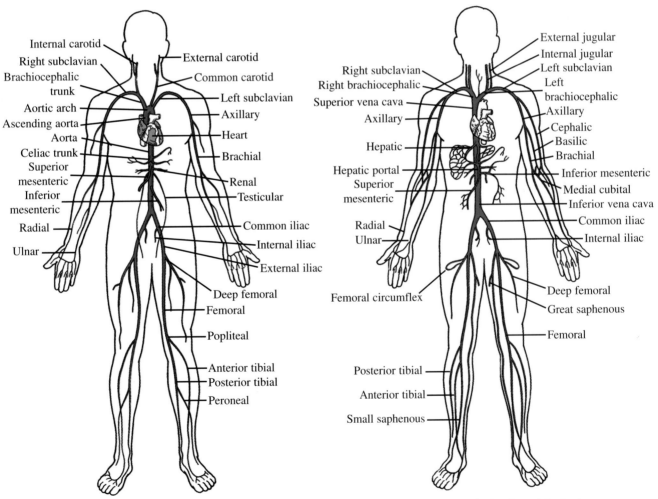

Figure 12.1 Principal arteries of the body. **Figure 12.2** Principal veins of the body.

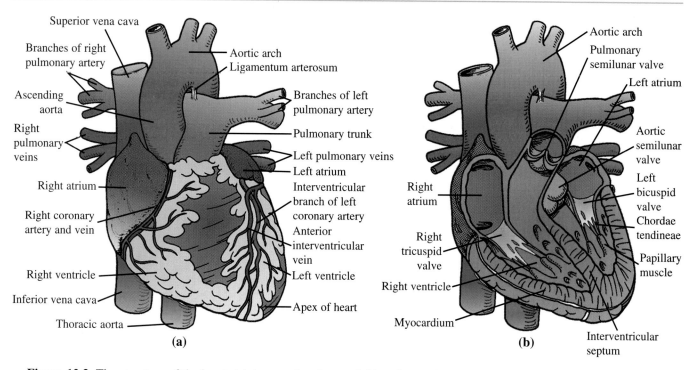

Figure 12.3 The structure of the heart. (a) An anterior view and (b) an internal view.

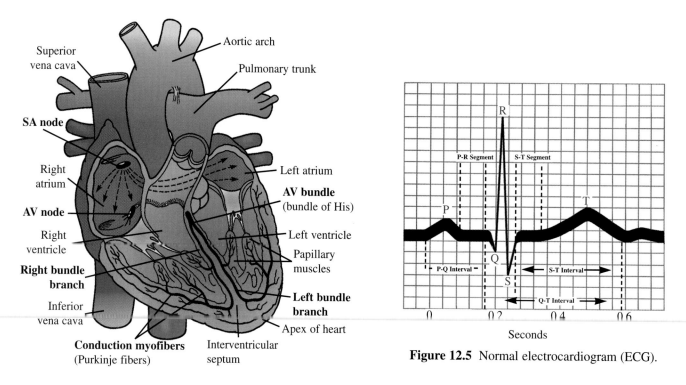

Figure 12.4 Conduction system of the heart, indicated in bold type.

Figure 12.5 Normal electrocardiogram (ECG).

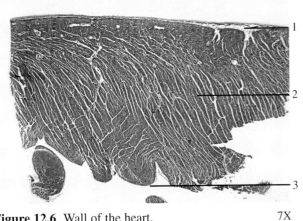
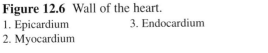

Figure 12.6 Wall of the heart.   7X

1. Epicardium   3. Endocardium
2. Myocardium

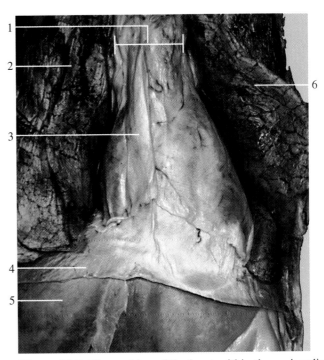
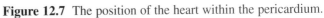

Figure 12.7 The position of the heart within the pericardium.

1. Mediastinum   4. Diaphragm
2. Right lung   5. Liver
3. Pericardium   6. Left lung

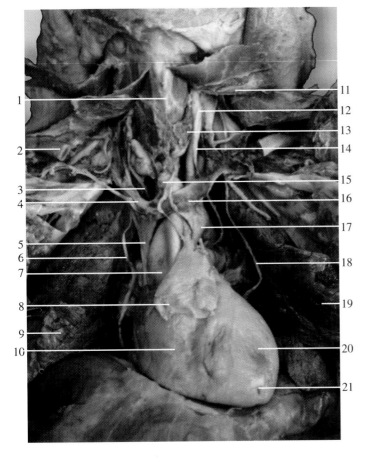

Figure 12.8 An antreior view of the heart and associated structures

1. Thyroid cartilage of larynx
2. First rib (cut)
3. Right vagus nerve
4. Right brachiocephalic vein
5. Superior vena cava
6. Right phrenic nerve
7. Ascending aorta
8. Pericardium (cut)
9. Right lung
10. Right ventricle of heart
11. Sternohyoid (cut and reflected)
12. Left common carotid artery
13. Thyroid gland
14. Left vagus nerve
15. Brachiocephalic artery
16. Left brachiocephalic artery
17. Aortic arch
18. Left phrenic nerve
19. Left lung
20. Left ventricle of heart
21. Apex of heart

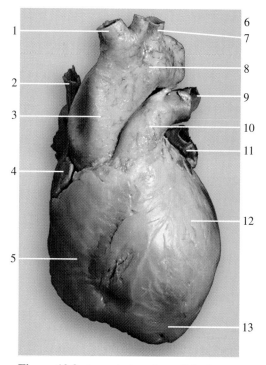

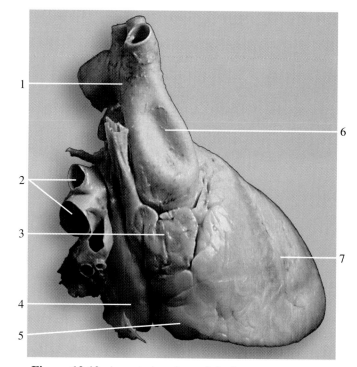

Figure 12.9 An anterior view of the heart and great vessels.

1. Brachiocephalic trunk
2. Superior vena cava
3. Ascending aorta
4. Right atrium
5. Right ventricle
6. Left common carotid artery
7. Left subclavian artery
8. Aortic arch
9. Pulmonary artery
10. Pulmonary trunk
11. Left atrium
12. Left ventricle
13. Apex of heart

Figure 12.10 A posterior view of the heart.

1. Aortic arch
2. Pulmonary arteries
3. Right atrium
4. Inferior vena cava
5. Right ventricle
6. Ascending aorta
7. Left ventricle

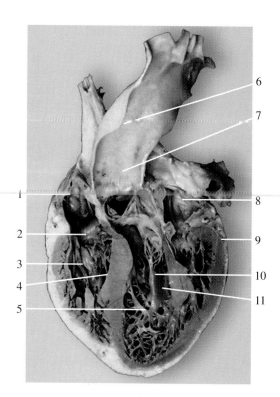

Figure 12.11 The internal structure of the heart.

1. Right atrium
2. Tricuspid valve
3. Right ventricle
4. Interventricular septum
5. Trabeculae carneae
6. Ascending aorta
7. Aortic semilunar valve
8. Bicuspid valve
9. Myocardium
10. Papillary muscle
11. Left ventricle

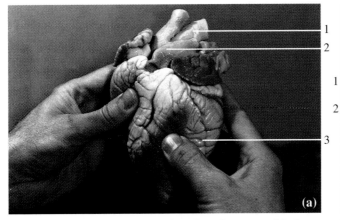

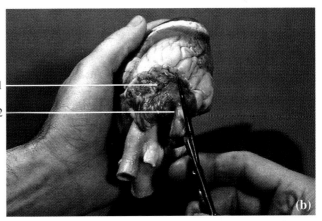

Position the heart so the ventral surface faces you. Notice the thicker ventricular walls, especially the left ventricle.
1. Aortic arch
2. Pulmonary trunk
3. Left ventricle

Insert the scissors into the superior vena cava. The cut should expose the interior of the right atrium. Notice the right tricuspid (atrioventricular) valve.
1. Right atrium
2. Superior vena cava

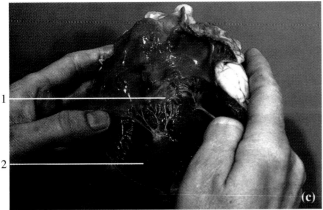

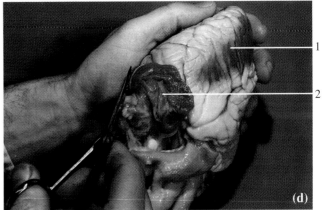

Continue the incision through the right ventricle to the apex of the heart. Observe the structure of the valve.
1. Right tricuspid valve
2. Right ventricle

Begin the next incision in the left atrium. This time, continue through both the atrium and the ventricle.
1. Left ventricle
2. Left atrium

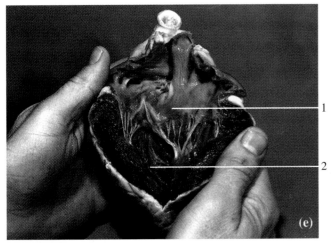

Expose the left ventricle and atrium. Notice the diffeence between the right and left ventricles, especially the thicker muscular wall of the left ventricle.

Figure 12.12 Sheep heart dissection.

1. Left bicuspid (mitral) valve.
2. Left ventricle

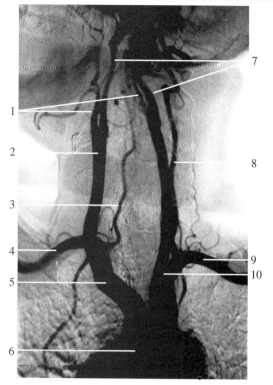

Figure 12.13 An angiogram showing the aortic arch and its branches.

1. External carotid arteries
2. Right common carotid artery
3. Right vertebral artery
4. Right subclavian artery
5. Brachiocephalic trunk
6. Aortic arch
7. Internal carotid arteries
8. Left vertebral artery
9. Left subclavian artery
10. Left common carotid artery

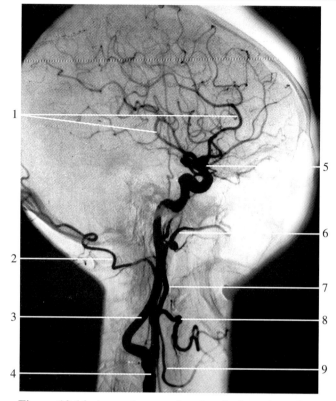

Figure 12.14 An angiogram showing the branches of the common carotid and external carotid arteries.

1. Meningeal arteries
2. Occipital artery
3. Internal carotid artery
4. Common carotid artery
5. Internal carotid artery to cerebral arterial circle (circle of Willis)
6. Maxillary artery
7. External carotid artery
8. Facial artery
9. Superior thyroid artery

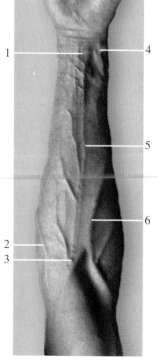

Figure 12.15
Surface anatomy identifying the superficial veins of the forearm.

1. Tendon of palmaris longus m.
2. Basilic vein
3. Median cubital vein
4. Site for palpation of radial artery
5. Median antebrachial vein
6. Cephalic vein

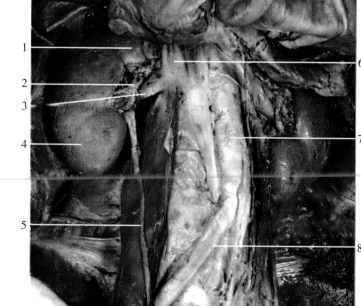

Figure 12.16 Arteries of the pelvic cavity.

1. Adrenal gland
2. Renal artery
3. Renal vein
4. Right kidney
5. Ureter
6. Inferior vena cava
7. Abdominal aorta
8. Right common iliac artery

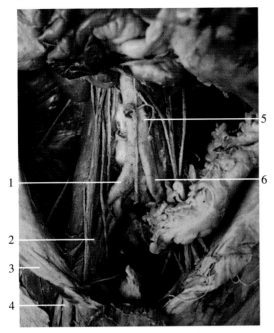

Figure 12.17 Arteries of the pelvic cavity.

1. Right common iliac artery
2. External iliac artery
3. Inguinal ligament
4. Femoral artery
5. Abdominal aorta
6. Left common iliac artery

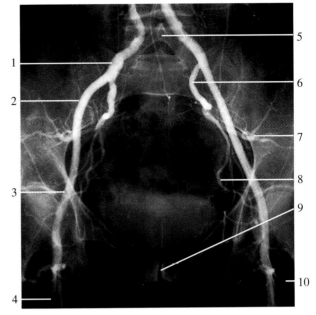

Figure 12.18 An angiogram of the common iliac arteries and their branches.

1. Common iliac artery
2. External iliac artery
3. Femoral artery
4. Deep femoral artery
5. Lumbar vertebra
6. Internal iliac artery
7. Gluteal arteries
8. Obturator artery
9. Symphysis pubis
10. Lateral circumflex femoral artery

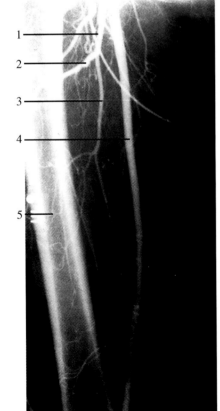

Figure 12.19 An angiogram of the arteries of the right thigh.

1. Deep femoral artery
2. Lateral circumflex femoral artery
3. Medial femoral circumflex artery
4. Femoral artery
5. Femur

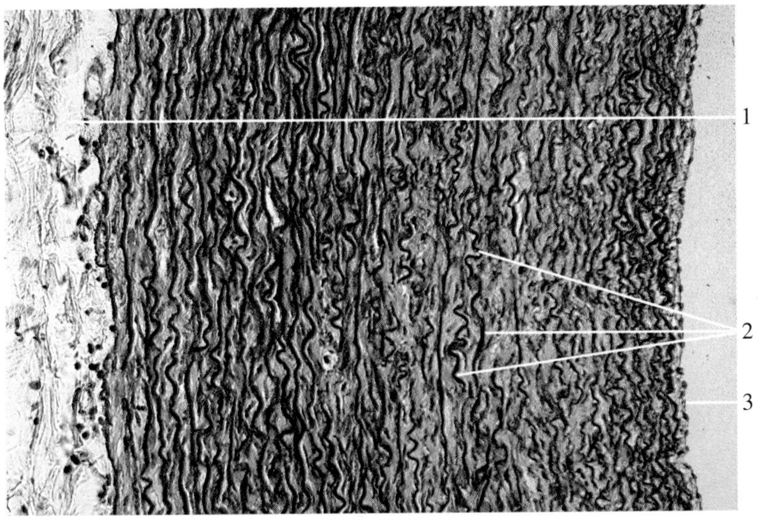

Figure 12.20 Wall of elastic artery. 200X

1. Tunica adventitia
2. Elastic laminae (in tunica media)
3. Tunica intima

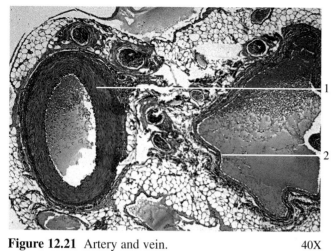

Figure 12.21 Artery and vein.   40X
1. Artery
2. Vein

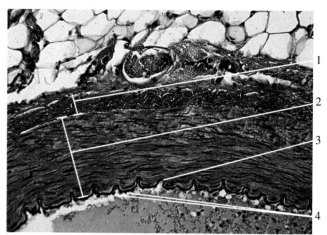

Figure 12.22 Wall of muscular artery.   75X
1. Tunica adventitia   3. Internal elastic membrane
2. Tunica media   4. Endothelial cells

Photo courtesy of Scott Miller

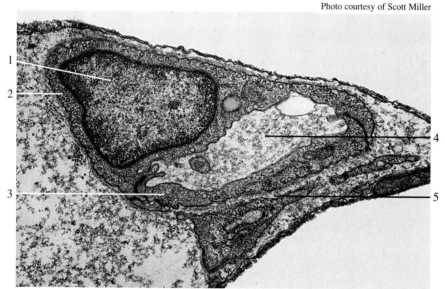

Figure 12.23 SEM photomicrograph of a capillary.
1. Nucleus
2. Endocytic vesicles
3. Endothelial cell
4. Lumen of capillary
5. Basal lamina

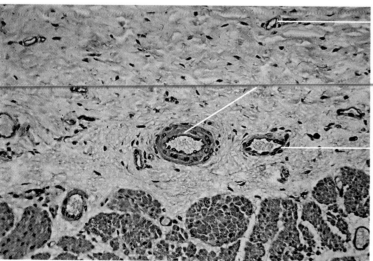

Figure 12.24 Arteriole, capillary, and venule.
1. Capillary
2. Arteriole
3. Venule

200X

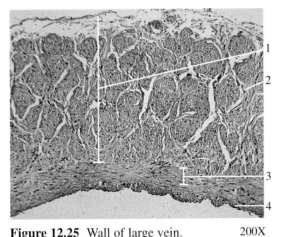

Figure 12.25   Wall of large vein.   200X
1. Tunica adventita
2. Longitudinally oriented smooth muscle
3. Tunica media
4. Tunica intima

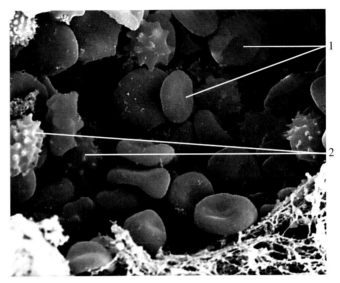

Figure 12.26   Electron micrograph of blood cells in the lumen of a blood vessel.
1. Erythrocytes   2. Leukocytes

Photo courtesy of Clifford E. Keeney

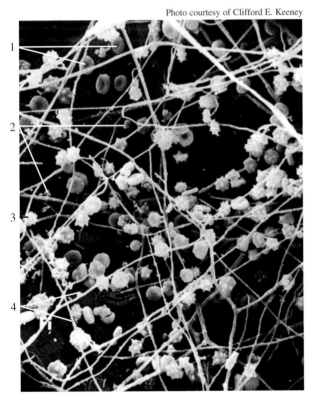

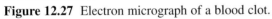

Figure 12.27   Electron micrograph of a blood clot.
1. Erythrocytes   3. Leukocyte
2. Thrombocytes   4. Fibrin strand

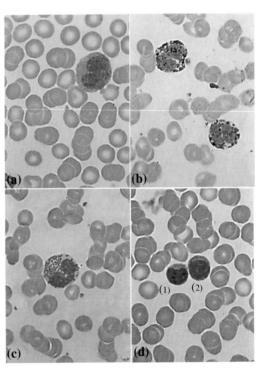

Figure 12.28   Types of leukocytes.   200X
(a) Neutrophil   (d) Monocyte(1);
(b) Basophils   Lymphocyte(2)
(c) Eosinophil

Lymphatic System

13

The lymphatic system is closely interrelated to the circulatory system. The functions of the lymphatic system are basically fourfold: 1) it transports excess interstitial (tissue) fluid, which was initially formed as a blood filtrate, back to the bloodstream; 2) it maintains homeostasis around body cells by providing a constantly moist intracellular environment, which assists movements of materials into and out of cells; 3) it serves as the route by which absorbed fat from the small intestine is transported to the blood; and 4) it helps provide immunological defenses against disease-causing agents.

Lymph capillaries drain tissue fluid, which is formed from blood plasma; when this fluid enters lymph capillaries, it is called *lymph*. Lymph is returned to the venous system via two large lymph ducts—the *thoracic duct* and the *right lymphatic duct* (fig. 13.1). On the way to these drainage ducts, lymph filters through *lymph nodes*, which contain phagocytic cells and germinal centers that produce lymphocytes. The *spleen* and *thymus* are considered lymphoid organs because they also produce lymphocytes.

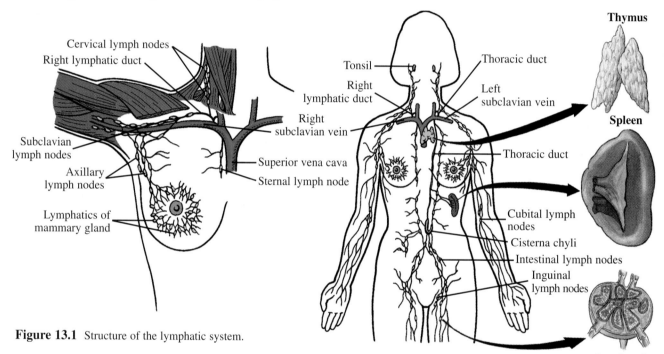

Figure 13.1 Structure of the lymphatic system.

Figure 13.2 Thymus.   7X
1. Cortex
2. Medulla

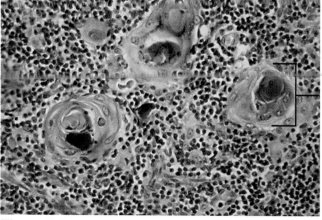

Figure 13.3 Thymic medulla.   300X
1. Hassall's corpuscle

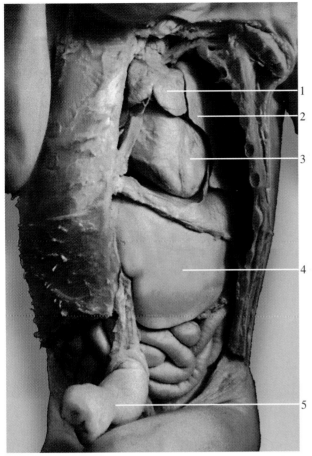

Figure 13. 4 The thymus within a fetus, during the third trimester of development.

1. Thymus   4. Liver
2. Lung   5. Umbilical cord
3. Heart

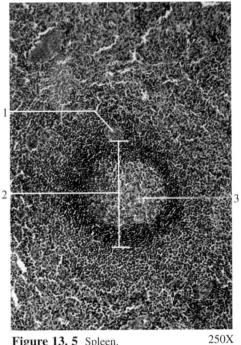

Figure 13. 5 Spleen.   250X

1. Central artery
2. Splenic nodule
3. Germinal center

Figure 13. 6 The spleen and pancreas.

1. Spleen
2. Splenic artery
3. Splenic vein
4. Pancreas
5. Pancreatic duct

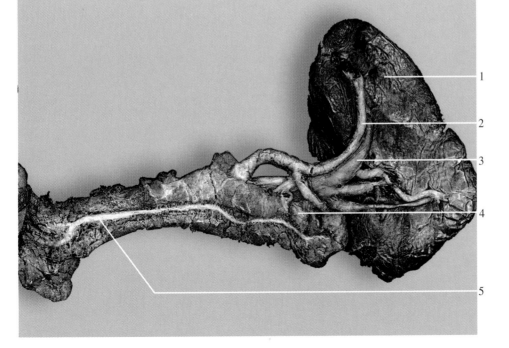

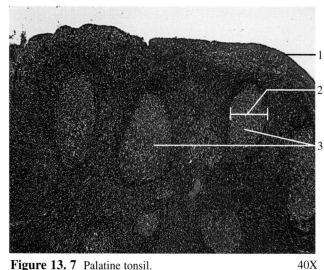

Figure 13. 7 Palatine tonsil.   40X
1. Oral mucosa
2. Lymphatic nodule
3. Germinal centers

Figure 13.8 Palatine tonsils that have been
removed in a tonsillectomy and sectioned by a
pathologist. Chronic tonsillitis generally
requires a tonsillectomy.

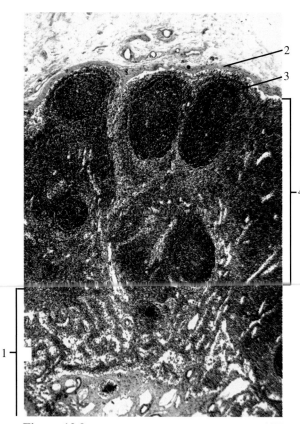

Figure 13.9 Lymph node.   40X
1. Medulla of lymph node   3. Lymphatic nodule
2. Capsule   4. Cortex of lymph node

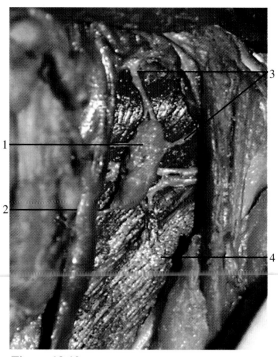

Figure 13.10 A lymph node.
1. Lymph node
2. Vein
3. Lymphatic vessels
4. Muscle

Respiratory System

14

The respiratory system is made up of organs and structures that function together to bring gases in contact with the blood of the circulatory system. This system consists of the *nasal cavity, pharynx, larynx*, and *trachea*, and the *bronchi, bronchioles*, and *pulmonary alveoli* within the lungs (fig. 14.2). The functions of the respiratory system are gas exchange, sound production, assistance in abdominal compression, and coughing and sneezing.

The nasal cavity has a bony and cartilaginous support. The ciliated, mucous lining of the upper respiratory tract warms, moistens, and cleanses inspired air. The *paranasal sinuses* are found in the maxillary, frontal, sphenoid, and ethmoid bones. The *pharynx* is a funnel-shaped passageway that connects the oral and nasal cavities with the larynx. The cartilaginous *larynx* keeps the passageway to the trachea open during breathing and closes the respiratory passageway during swallowing. It also contains the *vocal cords*. The *trachea* is a rigid tube, supported by rings of cartilage, that leads from the larynx to the *bronchial tree*. *Pulmonary alveoli* are the functional units of the lungs where gas exchange occurs; they are small, numerous, thin-walled air sacs. The right and left *lungs* are separated by the *mediastinum*; each lung is divided into *lobes* and *lobules*.

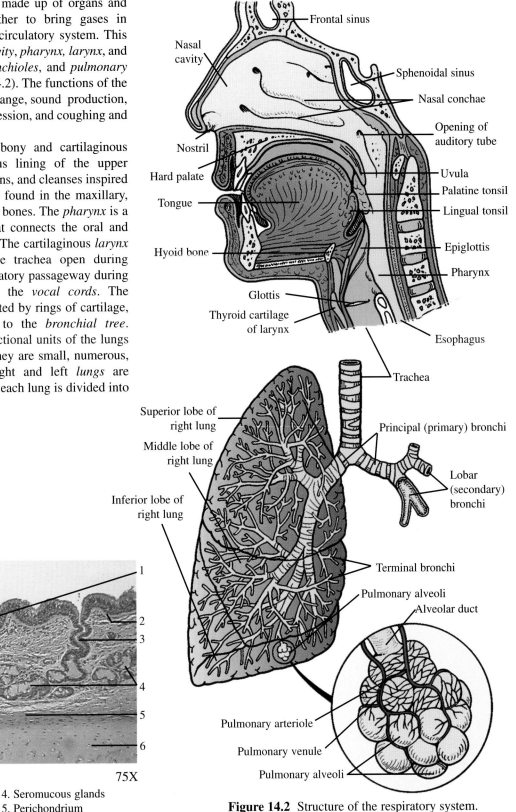

Figure 14.2 Structure of the respiratory system.

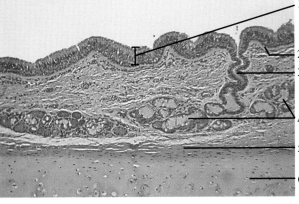

Figure 14.1 Tracheal wall. 75X
1. Respiratory epithelium
2. Basement membrane
3. Duct of seromucous gland
4. Seromucous glands
5. Perichondrium
6. Hyaline cartilage

113

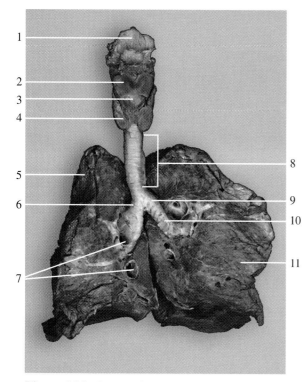

Figure 14.3 An anterior view of the larynx, trachea, and lungs.

1. Epiglottis
2. Thyroid cartilage
3. Cricoid cartilage
4. Thyroid gland
5. Right lung
6. Right principal (primary) bronchus
7. Pulmonary vessels
8. Trachea
9. Carnia
10. Left principal (primary) bronchus
11. Left lung

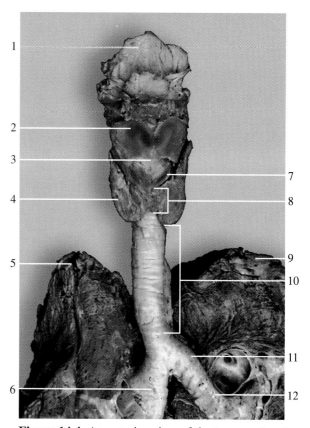

Figure 14.4 An anterior view of the larynx and trachea.

1. Epiglottis
2. Thyroid cartilage
3. Cricothyroid ligament
4. Thyroid gland
5. Superior lobe of right lung
6. Right principal (primary) bronchus
7. Cricoid cartilage
8. Isthmus of thyroid gland
9. Superior lobe of left lung
10. Trachea
11. Carnia
12. Left principal (primary) bronchus

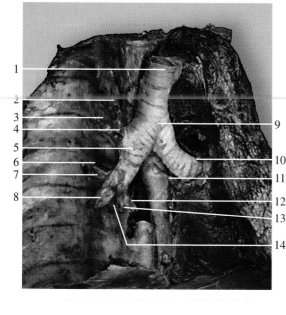

Figure 14.5 An anterior view of bronchi.

1. Trachea
2. Apical segmental bronchus
3. Posterior segmental bronchus
4. Anterior segmental bronchus
5. Right principal bronchus
6. Lateral segmental bronchus
7. Medial segmental bronchus
8. Anterior basal segmental bronchus
9. Carina
10. Left principal bronchus
11. Esophagus
12. Medial basal segmental bronchus
13. Posterior basal segmental bronchus
14. Lateral segmental bronchus

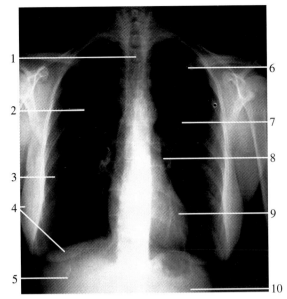

Figure 14.6 Radiograph of the thorax.

1. Thoracic vertebra
2. Right lung
3. Rib
4. Image of right breast
5. Diaphragm/liver

6. Clavicle
7. Left lung
8. Mediastinum
9. Heart
10. Diaphragm/stomach

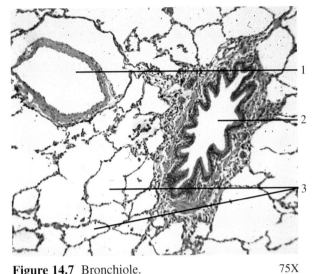

Figure 14.7 Bronchiole.   75X

1. Pulmonary arteriole   3. Pulmonary alveoli
2. Bronchiole

Figure 14.8 Electron micrograph of the lining of the trachea.

1. Cilia
2. Goblet cell

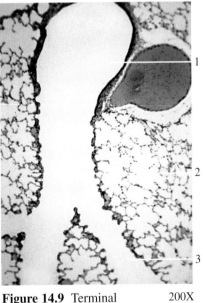

Figure 14.9 Terminal   200X
bronchiole.

1. Terminal bronchiole
2. Respiratory bronchiole
3. Alveolar duct

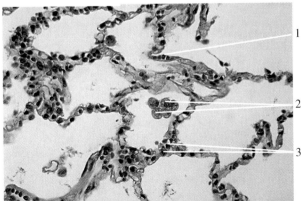

Figure 14.10 Pulmonary alveoli.

1. Capillary in alveolar wall
2. Macrophages
3. Type II pneumocytes

300X

Digestive System

15

The digestive system consists of a *gastrointestinal tract* (GI tract) and *accessory digestive organs*. Most of the food we eat is not suitable for cellular utilization until it is mechanically and chemically reduced to forms that can be absorbed through the intestinal wall and transported to the cells by the blood or lymph. Ingested food is not technically in the body until it is absorbed and, in fact, a large portion of consumed food is not digested at all but rather passes through as fecal material.

The functions of the principal regions and organs of the digestive system are presented in table 15.1. The digestive system is diagrammed in figure 15.1.

Table 15.1 Regions and Structures of the Digestive System

	Region or Structure	Function
Gastrointestinal tract	Oral cavity	Ingests food; receives saliva and initiates digestion of carbohydrates; mastication (chewing); forms bolus (food mass); deglutition (swallowing)
	Pharynx	Receives bolus from oral cavity and passes it to esophagus
	Esophagus	Transports bolus to stomach by peristalsis
	Stomach	Receives bolus from esophagus; forms chyme (paste-like food) initiates digestion of proteins; moves chyme into duodenum
	Small intestine	Receives chyme form stomach, along with secretions from liver and pancreas; chemically and mechanically breaks down chyme; absorbs nutrients; transports wastes to large intestine
	Large intestine	Receives undigested wastes from small intestine; absorbs water and electrolytes; forms and stores feces, and expels them through defecation
Accessory digestive organs	Teeth	Mechanically pulverize food
	Tongue	Manipulate food and assists in swallowing
	Salivary glands	Secrete saliva which aids in formation of bolus; initiates digestion of carbohydrates
	Liver	Production of bile; storage of iron and copper; conversion of glucose to glycogen and storage of glycogen; synthesis of certain vitamins; production of urea; synthesis of fibrinogen and prothrombin used for clotting of blood; phagocytosis of foreign material in blood; detoxifies harmful substances in body; storage of blood cells; hemopoiesis in fetus and newborn
	Gallbladder	Concentration and storage of bile necessary for emulsification of fats
	Pancreas	Production and secretion of pancreatic juice containing digestive enzymes; production and secretion of the hormones insulin and glucagon

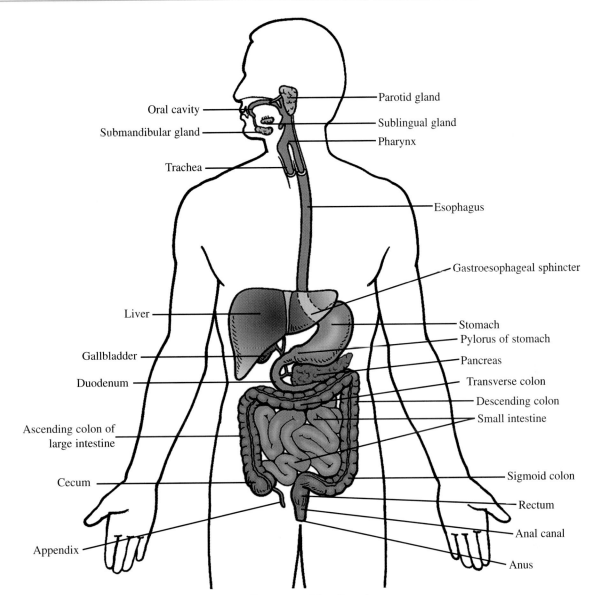

Figure 15.1 Structure of the digestive system.

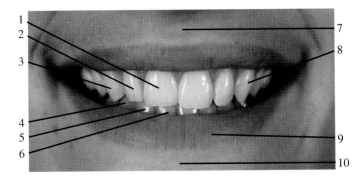

Figure 15.2 Oral region, lips, and teeth.

1. Medial incisor
2. Lateral incisor
3. Canine
4. Canine
5. Lateral incisor
6. Medial incisor
7. Philtrum
8. Canine
9. Inferior lip
10. Mentolabial sulcus

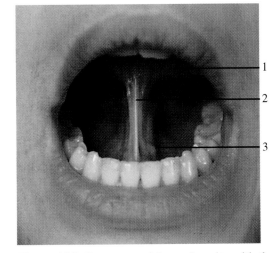

Figure 15.3 Structures of the oral cavity with the mouth open and the tongue elevated.

1. Tongue
2. Lingual frenulum
3. Opening of submandibular duct

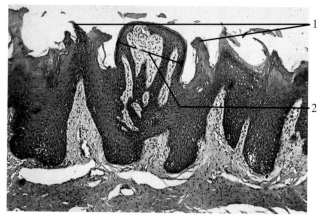

Figure 15.4 Filiform and fungiform papillae.   75X
1. Filiform papillae   2. Fungiform papilla

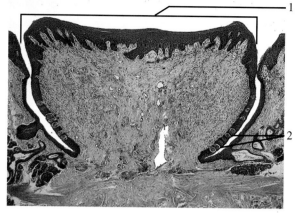

Figure 15.5 Vallate papilla   40X
1. Vallate papilla   2. Taste buds

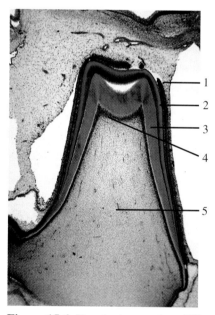

Figure 15.6 Developing tooth.   40X
1. Ameoblasts   4. Odontoblasts
2. Enamel   5. Pulp
3. Dentin

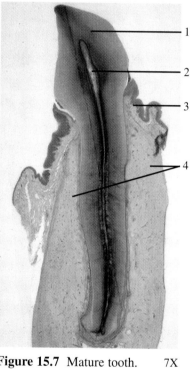

Figure 15.7 Mature tooth.   7X
1. Dentin (enamel has been
   dissolved away)
2. Pulp
3. Gingiva
4. Alveolar bone

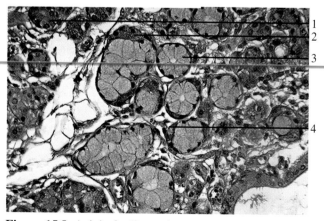

Figure 15.8 Acini of salivary tissue.   250X
1. Serous acinus   3. Mucous acinus
2. Serous demilune on   4. Serous demilune on
   mucous acinus   mucous acinus

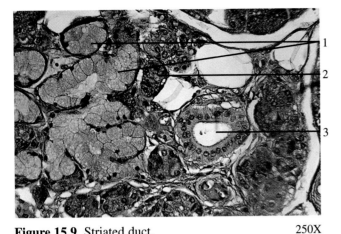

Figure 15.9 Striated duct.   250X
1. Mucous acini   3. Striated duct
2. Serous acinus

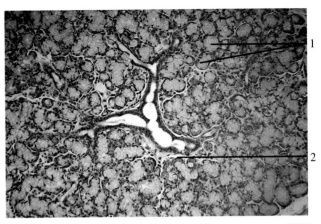

Figure 15.10 Sublingual gland (mostly mucous,   100X
some serous).
1. Mucous acini   2. Serous demilune

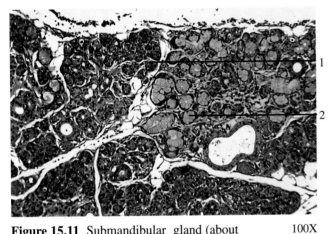

Figure 15.11 Submandibular gland (about   100X
one-half mucous and one-half serous).
1. Serous acinus   2. Mucous acinus

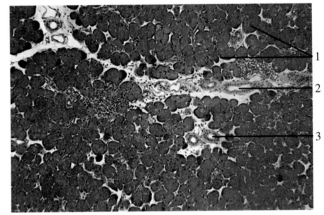

Figure 15.12 Parotid gland (purely serous).   250X
1. Serous acini   3. Striated duct
2. Excretory duct

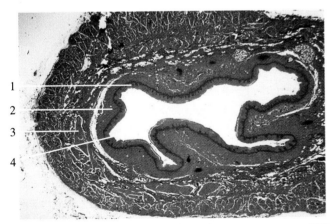

Figure 15.13 Cross section of esophagus.   10X
1. Mucosa   3. Muscularis
2. Submucosa   4. Lumen

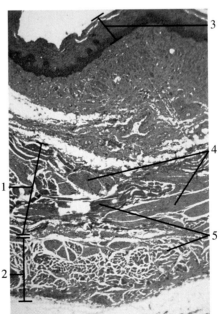

1. Inner circular layer (muscularis externa)
2. Outer longitudinal layer (muscularis externa)
3. Epithelium
4. Smooth muscle
5. Skeletal muscle

Figure 15.14 Wall of esophagus.   30X

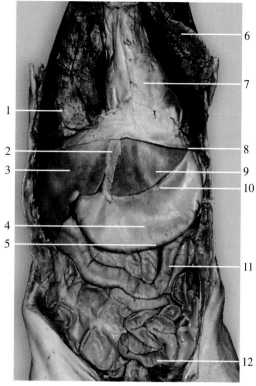

Figure 15.15 An anterior aspect of the trunk.
1. Right lung
2. Falciform ligament
3. Right lobe of liver
4. Body of stomach
5. Greater curvature of stomach
6. Left lung (reflected)
7. Pericardium
8. Diaphragm
9. Left lobe of liver
10. Lesser curvature of stomach
11. Transverse colon
12. Small intestine

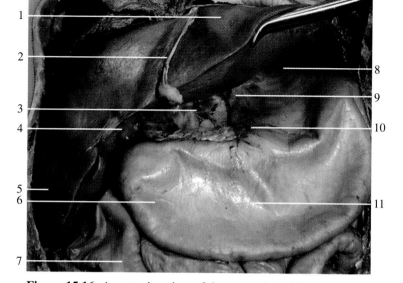

Figure 15.16 An anterior view of the stomach and liver.
1. Left lobe of liver
2. Falciform ligament
3. Celiac trunk (covered by hepatoduodenal ligament)
4. Gallbladder
5. Right lobe of liver
6. Pylorus of stomach
7. Transverse colon
8. Fundus of stomach
9. Lesser omentum
10. Gastric vein (traversing through omentum
11. Body of stomach

Figure 15.17 A CT scanogram of the trunk.
1. Right lung within pleural cavity
2. Thoracic vertebra
3. Liver
4. Small intestine
5. Ascending colon
6. Heart
7. Diaphragm
8. Stomach
9. Descending colon
10. Ilium

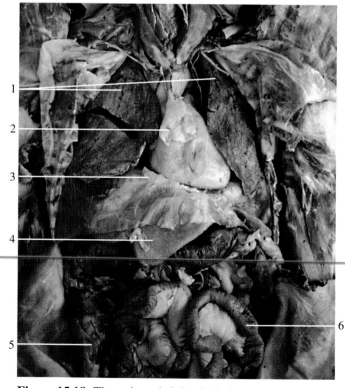

Figure 15.18 Thoracic and abdominal viscera.
1. Lungs
2. Heart (surrounded by pericardial fat)
3. Diaphragm
4. Liver
5. Cecum
6. Small intestine

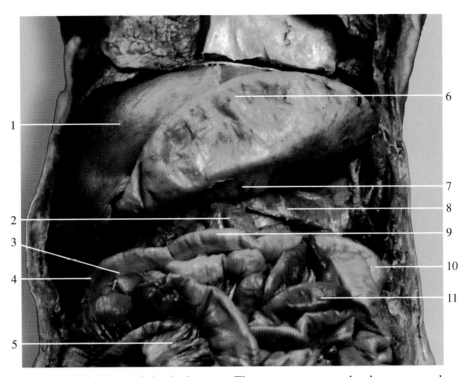

Figure 15.19 Upper abdominal organs. The greater omentum has been removed and the stomach reflected.

1. Right lobe of liver
2. Duodenum
3. Taeniae coli
4. Right colic (hepatic) flexure
5. Mesentery
6. Greater curvature of stomach

7. Lesser omentum
8. Pancreas
9. Transverse colon
10. Left colic (splenic) flexure
11. Small intestine

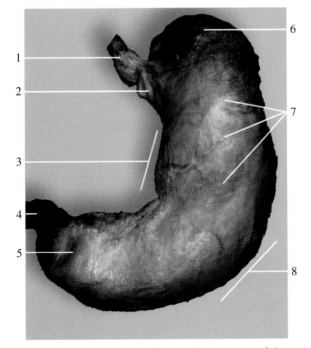

Figure 15.20 Major regions and structures of the stomach.

1. Esophagus
2. Cardiac portion of stomach
3. Lesser curvature of stomach
4. Duodenum
5. Pylorus of stomach
6. Fundus of stomach
7. Body of stomach
8. Greater curvature of stomach

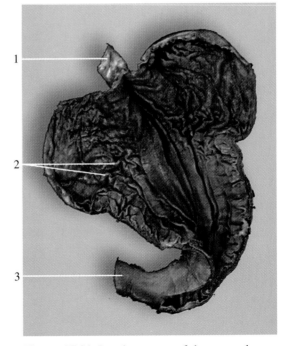

Figure 15.21 Interior aspect of the stomach.

1. Esophagus
2. Gastric folds (rugae)
3. Duodenum

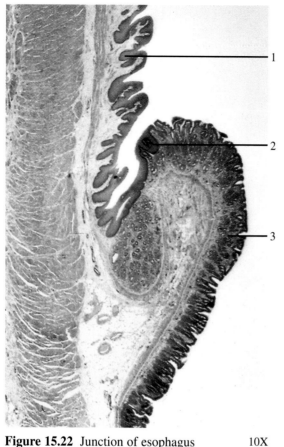

Figure 15.22 Junction of esophagus   10X
and stomach.
1. Epithelium of esophagus
2. Abrupt change in epithelium
3. Epithelium of stomach

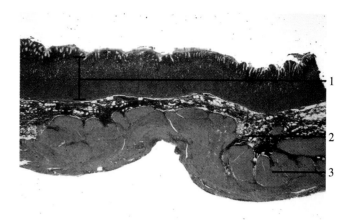

Figure 15.23 Wall of stomach.   10X
1. Mucosa   3. Muscularis externa
2. Submucosa

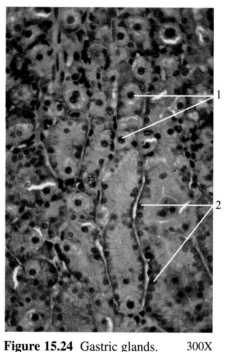

Figure 15.24 Gastric glands.   300X
1. Parietal cells
2. Chief cells

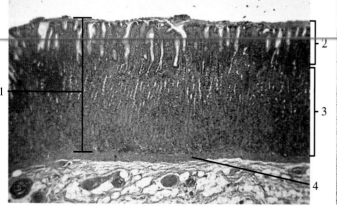

Figure 15.25 Mucosa of stomach body.   30X
1. Mucosa   3. Gastric glands
2. Gastric pits   4. Muscularis mucosae

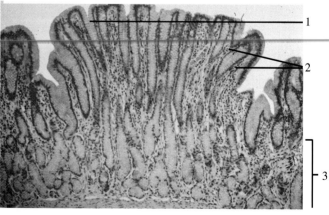

Figure 15.26 Mucosa of stomach pylorus.   200X
1. Gastric pit   3. Gastric glands
2. Lamina propria

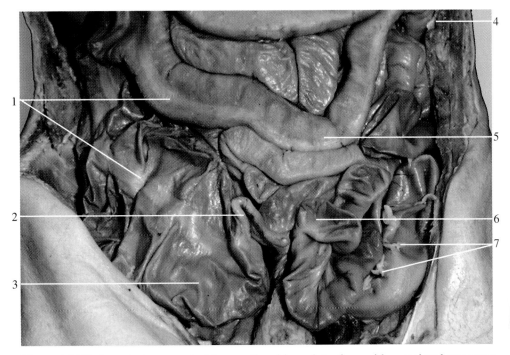

Figure 15.27 An anterior aspect of the small and large intestines with associated structures.

1. Taeniae coli
2. Appendix
3. Cecum
4. Left colic (splenic) flexure

5. Transverse colon
6. Small intestine
7. Epiploic appendages

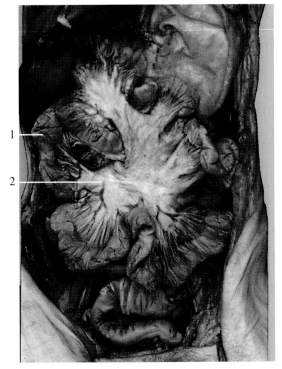

Figure 15.28 Mesentery and small intestine.
1. Serosa (visceral peritoneum)
2. Mesentery

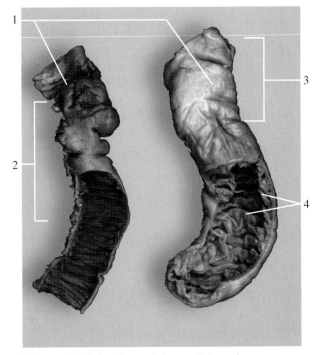

Figure 15.29 Sections of the small intestine.
1. Serosa (visceral peritoneum)
2. Ileum
3. Jejunum
4. Plicae circulares

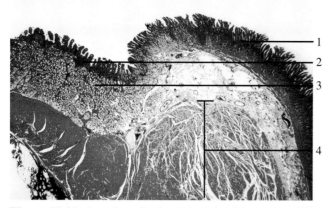

Figure 15.30 Junction of pylorus and duodenum. 10X
1. Stomach epithelium
2. Duodenal epithelium
3. Duodenal (Brunner's) glands
4. Pyloric sphincter

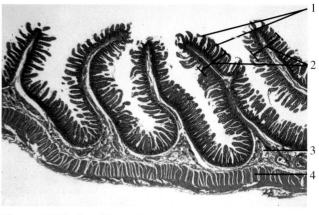

Figure 15.31 Small intestine. 7X
1. Villi
2. Plicae circulares
3. Submucosa
4. Muscularis externa

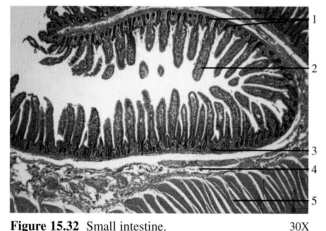

Figure 15.32 Small intestine. 30X
1. Intestinal glands
2. Villus
3. Muscularis mucosae
4. Submucosa
5. Muscularis externa

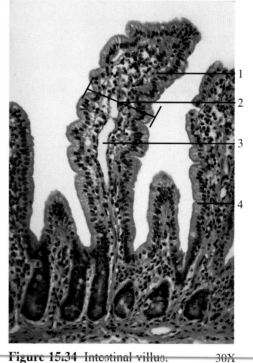

Figure 15.34 Intestinal villus. 30X
1. Lamina propria
2. Villus
3. Central lacteal
4. Simple columnar epithelium

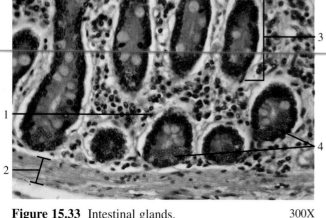

Figure 15.33 Intestinal glands. 300X
1. Eosinophil within lamina propria
2. Muscularis mucosae
3. Intestinal gland
4. Paneth cells

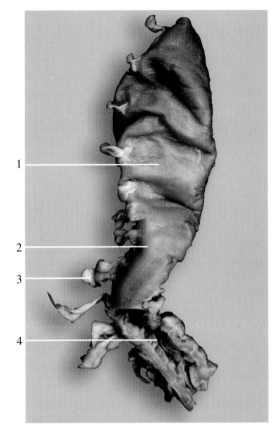

Figure 15.35 Section of the large intestine (colon).

1. Haustrum
2. Taeniae coli
3. Epiploic appendage
4. Semilunar folds (plicae) of colon

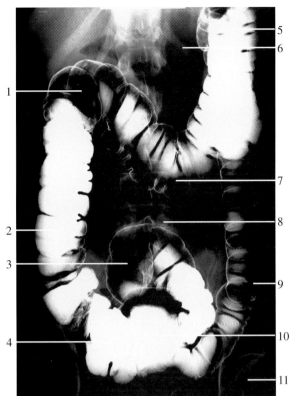

Figure 15.36 Radiograph of the large intestine.

1. Right colon (hepatic) flexure
2. Ascending colon
3. Sigmoid colon
4. Cecum
5. Left colic (splenic) flexure
6. 12th rib
7. Transverse colon
8. Lumbar vertebra
9. Descending colon
10. Rectum
11. Hip joint

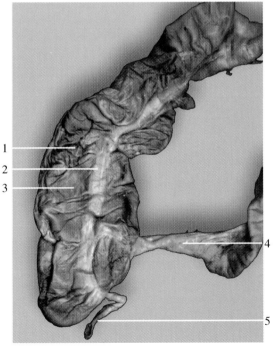

Figure 15.37 The cecum and appendix.

1. Ascending colon
2. Taenia coli
3. Cecum
4. Ileum
5. Appendix

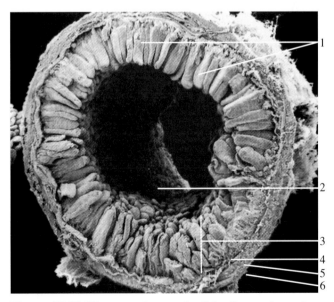

Figure 15.38 Electron micrograph of the ileum, shown in cross section.

1. Villi
2. Lumen
3. Mucosa
4. Submucosa
5. Tunica muscularis
6. Adventitia

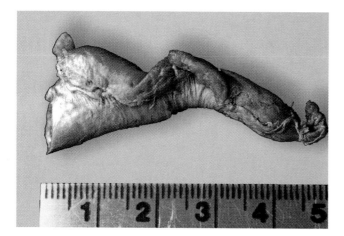

Figure 15.39 An inflamed appendix that has been removed in an appendectomy. The chief danger of appendicitis is the the appendix might rupture and produce peritonitis.

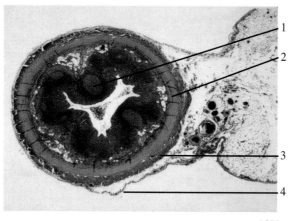

10X

Figure 15.40 Appendix.
1. Lymphatic nodule
2. Circular layer of muscularis externa
3. Longitudinal layer of muscularis externa
4. Serosa (peritoneum)

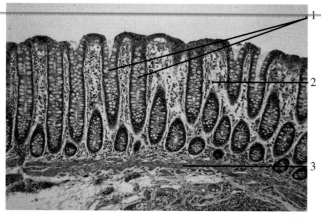

75X

Figure 15.41 Large intestine.
1. Glands
2. Lamina propria
3. Muscularis mucosae

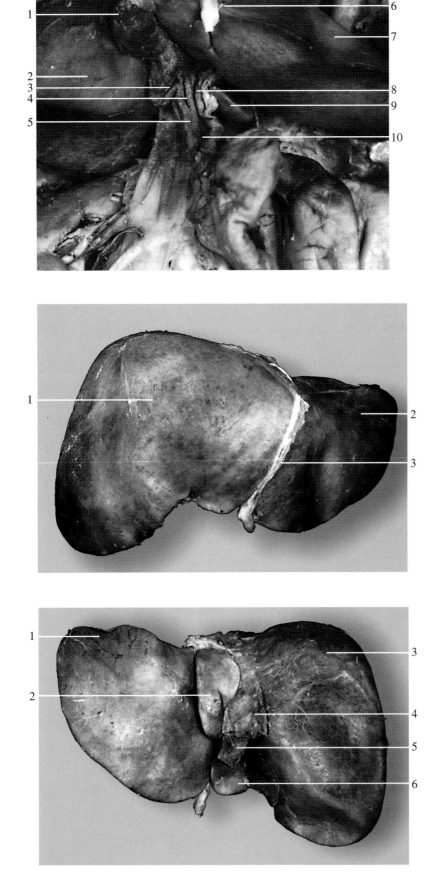

Figure 15.42 An inferior view of the liver and gallbladder.
1. Gallbladder
2. Right lobe of liver
3. Cystic duct
4. Common hepatic duct
5. Common bile duct
6. Falciform ligament
7. Left lobe of liver
8. Hepatic artery
9. Caudate lobe of liver
10. Hepatic portal vein

Figure 15.43 An anterior view of the liver.
1. Right lobe of liver
2. Left lobe of liver
3. Falciform ligament

Figure 15.44 An inferior view of the liver.
1. Left lobe of liver
2. Caudate lobe of liver
3. Right lobe of liver
4. Hepatic portal vein
5. Hepatic artery
6. Quadrate lobe of liver

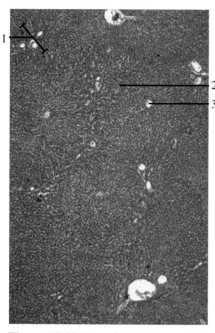

Figure 15.45 Hepatic lobules.   75X
1. Portal area   3. Central vein
2. Hepatic lobule

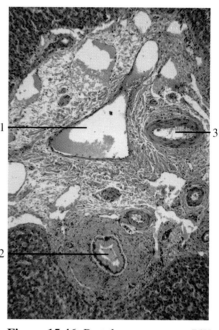

Figure 15.46 Portal area.   75X
1. Portal vein   3. Hepatic artery
2. Bile duct

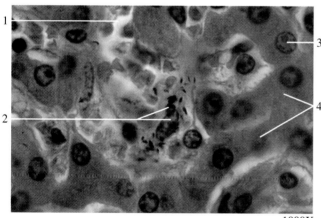
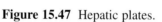

1000X

Figure 15.47 Hepatic plates.
1. Sinusoid containing red blood cells
2. Kupffer cell with debris in it's cytoplasm
3. Hepatocyte nucleus
4. Plate of hepatocytes

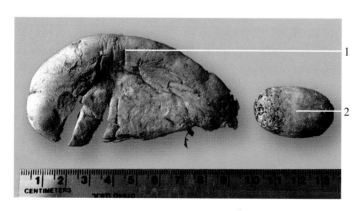

Figure 15.48 A gallbladder that has been
removed (cholecystectomy) and cut open to
remove it's gall stone (biliary calculus).
1. Gallbladder
2. Gall stone

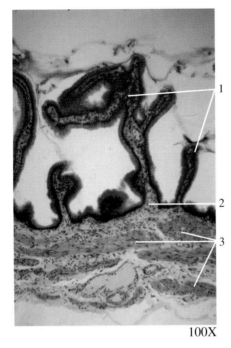

100X

Figure 15.49 Gallbladder.
1. Mucosal folds
2. Lamina propria
3. Muscularis

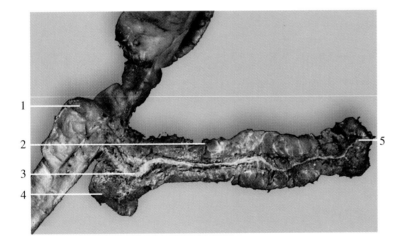

Figure 15.50 An anterior aspect of the pancreas and pancreatic duct.
1. Duodenum
2. Body of pancreas
3. Pancreatic duct
4. Head of pancreas
5. Tail of pancreas

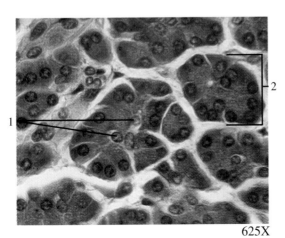

625X

Figure 15.51 Pancreatic acini.
1. Centroacinar cells
2. Acinus

Urinary System

16

The urinary system consists of the *kidneys, ureters, urinary bladder,* and *urethra* (fig. 16.1). The urinary system: 1) removes metabolic wastes from the blood and excretes it (as urine) to the outside during micturition (the physiological aspect of urination); 2) regulates, in part, the rate of red blood cell formation by secretion of the hormone *erythropoietin*; 3) assists the regulation of blood pressure by secreting the enzyme *renin*; 4) assists the regulation of calcium by activating vitamin D; and 5) assists the regulation of the volume, composition, and pH of body fluids.

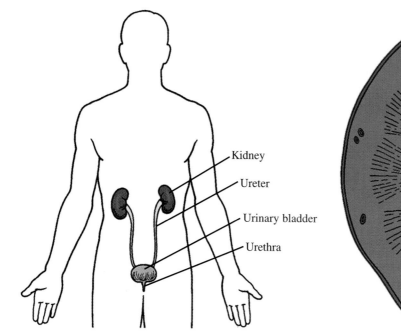

Figure 16.1 Organs of the urinary system.

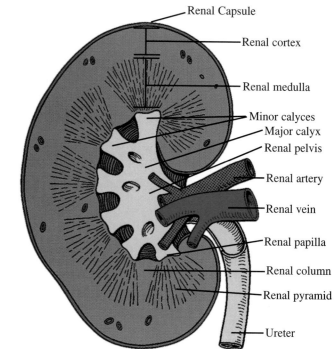

Figure 16.2 Structure of the kidney.

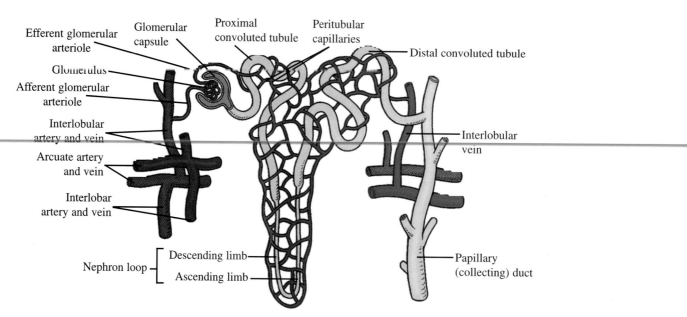

Figure 16.3 Structure of the nephron.

130

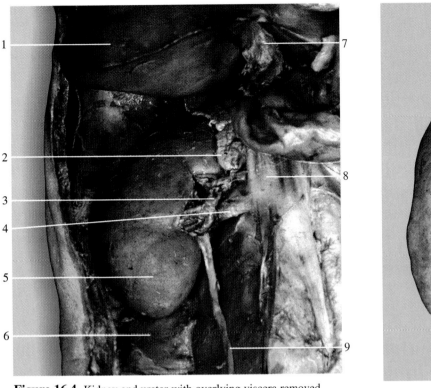

Figure 16.4 Kidney and ureter with overlying viscera removed.
1. Liver
2. Adrenal gland
3. Renal artery
4. Renal vein
5. Right kidney
6. Quadratus lumborum muscle
7. Gallbladder
8. Inferior vena cava
9. Ureter

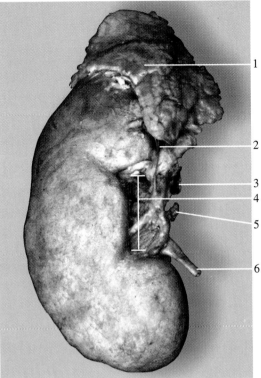

Figure 16.5 An anterior view of the right kidney.
1. Adrenal gland
2. Suprerenal artery
3. Renal artery
4. Hilum
5. Renal vein
6. Ureter

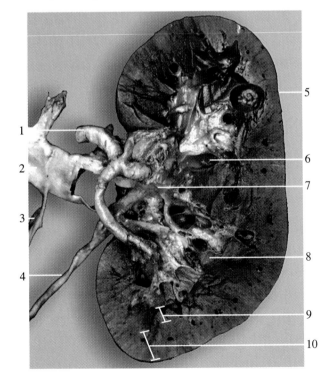

Figure 16.6 A coronal section of the left kidney.
1. Renal artery
2. Renal vein
3. Left testicular vein
4. Ureter
5. Capsule
6. Major calyx
7. Renal pelvis
8. Renal papilla
9. Renal medulla
10. Renal cortex

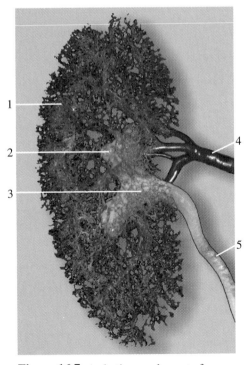

Figure 16.7 A plastic vascular cast of a kidney and ureter.
1. Arteriole network
2. Major calyx
3. Renal pelvis
4. Renal artery
5. Ureter

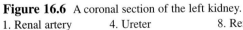

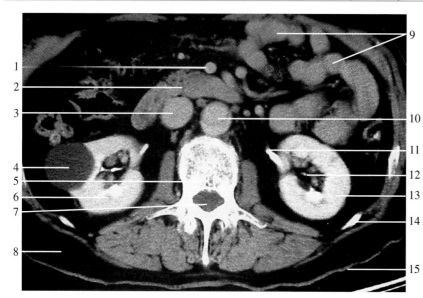

Figure 16.8 A transaxial CT image of abdominal viscera. Note the large renal cyst in the right kidney.

1. Superior mesenteric artery
2. Pancreas
3. Inferior vena cava
4. Renal crest
5. Body of lumbar vertebra
6. Right kidney
7. Vertebral foramen
8. Subcutaneous fat
9. Small intestine
10. Abdominal portion of aorta
11. Left ureter
12. Renal pelvis
13. Renal cortex
14. Rib
15. Skin

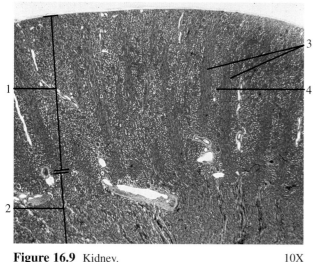

Figure 16.9 Kidney.   10X

1. Renal cortex
2. Renal medulla
3. Medullary rays
4. Cortical labyrinth

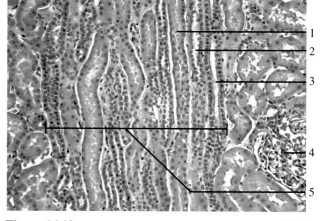

Figure 16.10 Medullary ray.   100X

1. Proximal tubule (of straight portion)
2. Distal tubule (of straight portion)
3. Collecting tubule
4. Glomerulus
5. Medullary ray

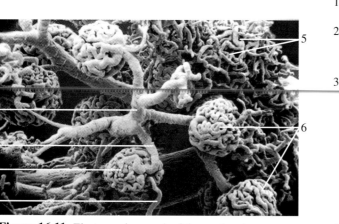

Figure 16.11 Electron micrograph of the renal cortex.

1. Interlobular artery
2. Afferent glomerular arteriole
3. Glomerulus
4. Efferent glomerular arteriole
5. Peritubular capillaries
6. Glomeruli

Figure 16.12 Renal corpuscle.   250X

1. Glomerular capsule
2. Urinary space
3. Glomerulus
4. Macula densa of distal tubule
5. Renal corpuscle

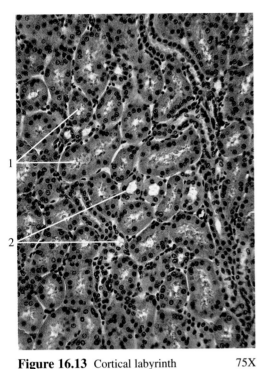

Figure 16.13 Cortical labyrinth   75X
1. Proximal convoluted tubule
2. Distal convoluted tubule

Figure 16.14 Renal papilla.   100X
1. Renal papilla
2. Minor calyx
3. Transitional epithelium

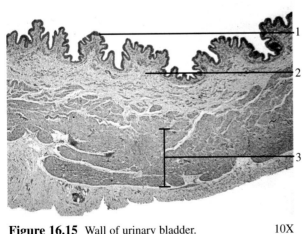

Figure 16.15 Wall of urinary bladder.   10X
1. Transitional epithelium   3. Muscularis
2. Lamina propria

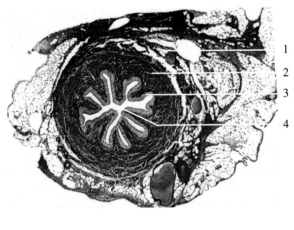

Figure 16.16 Ureter.   15X
1. Adventitia   3. Mucosa
2. Muscularis   4. Lumen

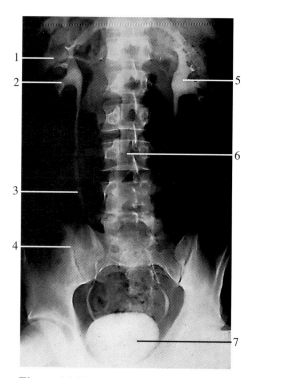

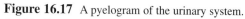

Figure 16.17 A pyelogram of the urinary system.
1. 12th rib   5. Renal pelvis
2. Major calyx   6. 3rd lumbar vertebra
3. Ureter   7. Urinary bladder
4. Sacroiliac joint

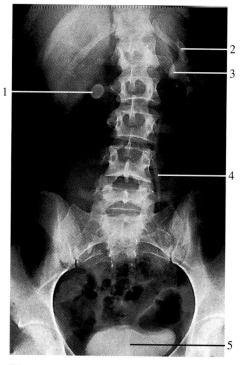

Figure 16.18 A radiograph showing a renal calculus (kidney stone) in the renal pelvis of the right kidney.
1. Renal calculus   4. Left ureter
2. Major calyx of left kidney   5. Urinary bladder
3. Renal pelvis of kidney

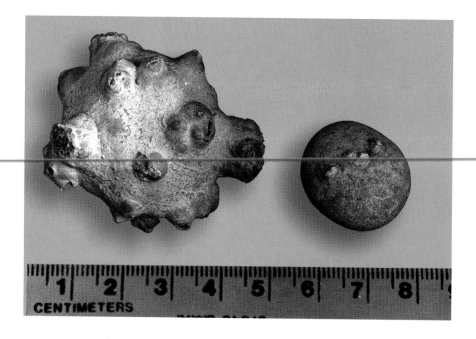

Figure 16.19 Renal calculi (stones) from the urinary bladders of two different patients. Renal calculi vary considerably in appearance and size and may develop in the renal pelvis of the kidney or in the urinary bladder.

Reproductive System

17

The organs of the male and female reproductive systems produce gametes and provide the mechanism for the union of these sex cells during the process of coitus (sexual intercourse). It is through sexual reproduction that individuals of a species are propagated, each having a genetic diversity inherited from both parents.

The male reproductive system consists of the *testes* (in the *scrotum*), excretory glands (*seminal vesicles*, *prostate*, and *bulbourethral glands*), and *penis*. The functions of the male reproductive system are to secrete sex hormones, produce spermatozoa, and ejaculate semen (spermatozoa and additives) into the vagina of the female.

The female reproductive system consists of the *ovaries*, *uterine tubes*, *vagina*, *external genitalia*, and *mammary glands*. The functions of the female reproductive system are to secrete sex hormones, produce ova, receive ejaculated semen from the erect penis of the male, nourish the developing embryo and fetus, deliver the baby, and nurse the infant once it is born.

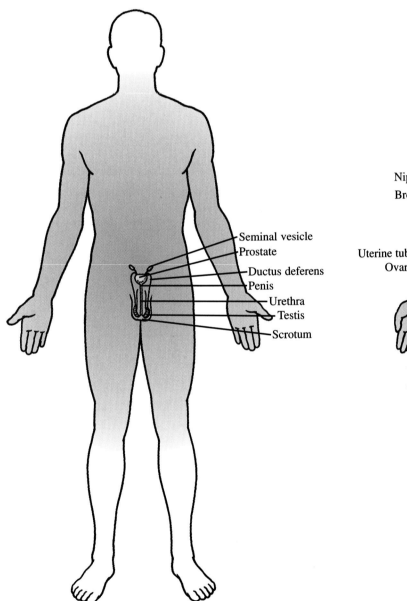

Figure 17.1 Organs of the male reproductive system.

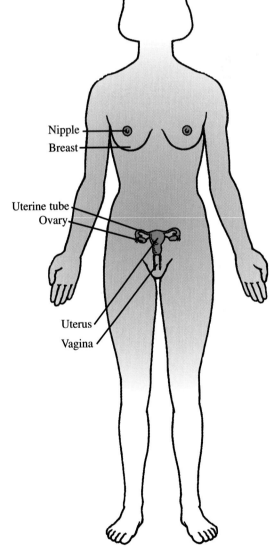

Figure 17.2 Organs of the female reproductive system.

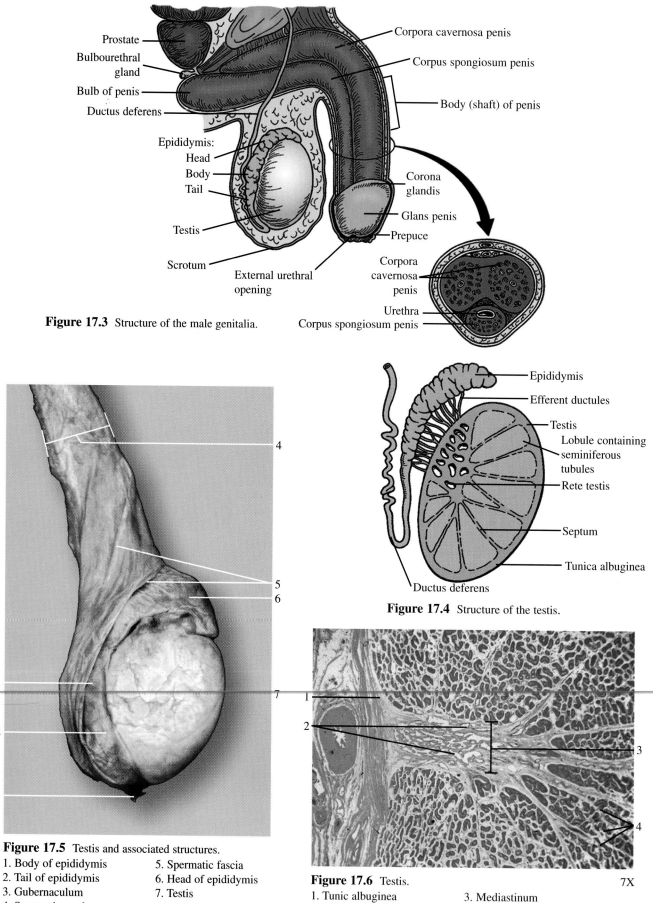

Prostate
Bulbourethral gland
Bulb of penis
Ductus deferens
Epididymis:
Head
Body
Tail
Testis
Scrotum
External urethral opening

Corpora cavernosa penis
Corpus spongiosum penis
Body (shaft) of penis
Corona glandis
Glans penis
Prepuce
Corpora cavernosa penis
Urethra
Corpus spongiosum penis

Figure 17.3 Structure of the male genitalia.

Epididymis
Efferent ductules
Testis
Lobule containing seminiferous tubules
Rete testis
Septum
Tunica albuginea
Ductus deferens

Figure 17.4 Structure of the testis.

Figure 17.5 Testis and associated structures.
1. Body of epididymis
2. Tail of epididymis
3. Gubernaculum
4. Spermatic cord
5. Spermatic fascia
6. Head of epididymis
7. Testis

Figure 17.6 Testis.   7X
1. Tunic albuginea
2. Tubules of rete testis
3. Mediastinum
4. Seminiferous tubules

Figure 17.7 Electron micrograph of a seminiferous tubule.
1. Spermatozoa   3. Spermatogonia
2. Spermatids

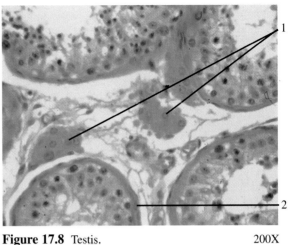

Figure 17.8 Testis.   200X
1. Interstitial (Leydig) cells
2. Seminiferous tubule

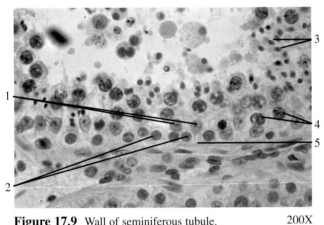

Figure 17.9 Wall of seminiferous tubule.   200X
1. Sustentacular (Sertoli) cells   4. Primary spermatocytes
2. Spermatogonia   5. Boundary of seminiferous
3. Spermatids   tubule

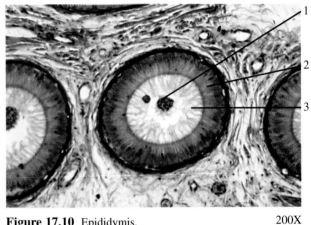

Figure 17.10 Epididymis.   200X
1. Sperm in lumen   3. Stereocilia
2. Pseudostratified columnar
   epithelium

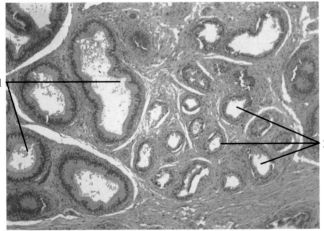

Figure 17.11 Efferent ductules.   75X
1. Duct of epididymis
2. Efferent ductules

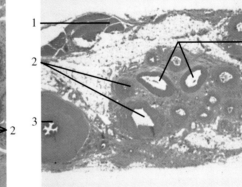

Figure 17.12 Spermatic cord.   10X
1. Cremaster muscle   3. Ductus deferens
2. Veins of the pampinform   4. Testicular artery
   plexus   5. Cremaster muscle

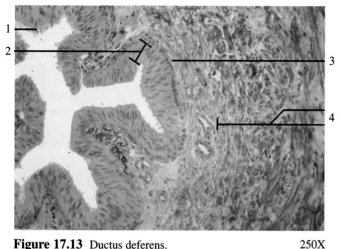

Figure 17.13 Ductus deferens.   250X
1. Stereocilia
2. Pseudostratified columnar epithelium
3. Lamina propria
4. Muscularis

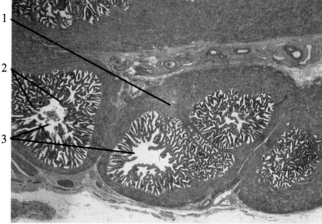

Figure 17.14 Seminal vesicle.   30X
1. Fibromuscular stroma
2. Mucosal folds
3. Lumen

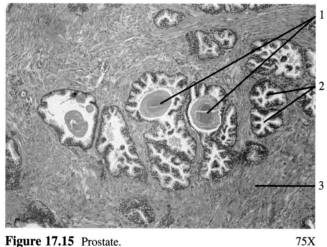

Figure 17.15 Prostate.   75X
1. Prostate concretions
2. Glandular acini
3. Fibromuscular stroma

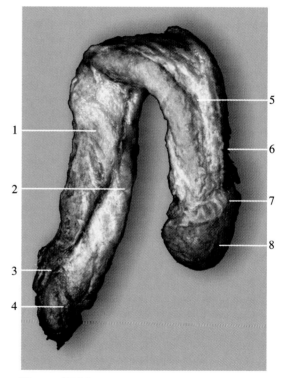

Figure 17.16 Structure of the penis.
1. Corpora cavernosa   5. Body of penis
2. Corpus spongiosum   6. Skin of penis
3. Crus of penis   7. Corona glandis
4. Bulb of penis   8. Glans penis

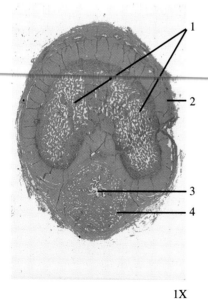

Figure 17.17 Penis.
1. Corpora cavernosa
2. Tunica albuginea
3. Urethra
4. Corpus spongiosum

1X

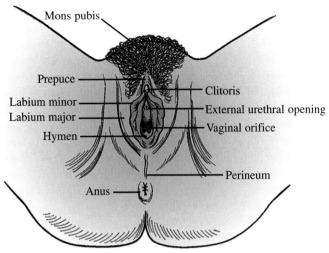

Figure 17.18 Female external genitalia.

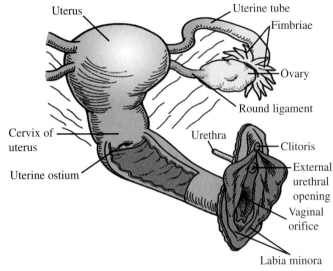

Figure 17.19 External genitalia and internal reproductive organs of the female reproductive system..

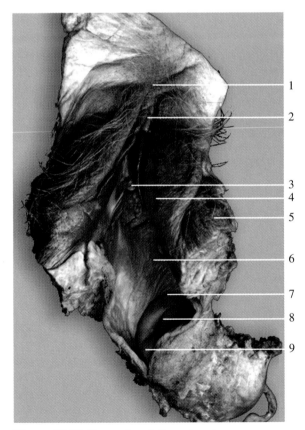

Figure 17.20 External genitalia and vagina.
1. Mons pubis
2. Clitoris
3. Urethral opening
4. Labium minor
5. Labium major
6. Vagina (dissected open)
7. Fornix
8. Uterine ostium
9. Cervix of uterus

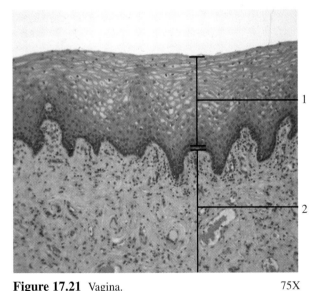

Figure 17.21 Vagina.   75X
1. Nonkeratinized stratified squamous epithelium
2. Lamina propria

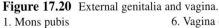

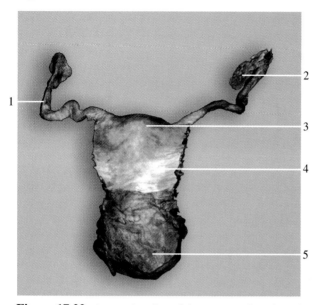

Figure 17.22 A posterior view of the uterus and uterine tubes.

1. Uterine tube   4. Body of uterus
2. Fimbriae   5. Cervix of uterus
3. Fundus of uterus

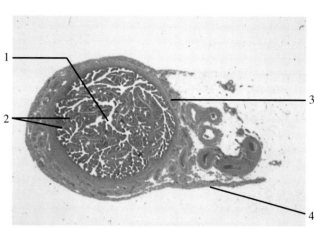

Figure 17.23 Ampulla of uterine tube.   10X

1. Lumen   3. Muscularis
2. Mucosal folds   4. Serosa

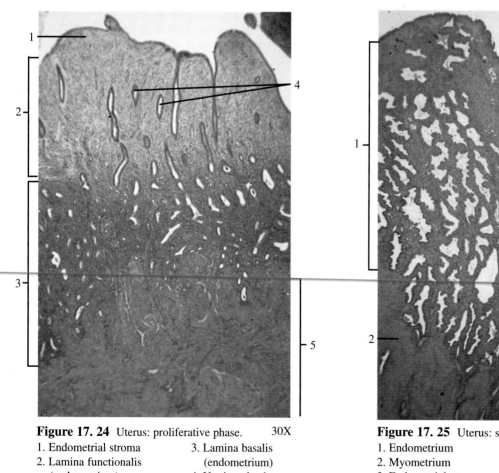

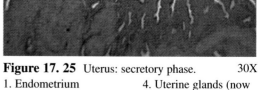

Figure 17. 24 Uterus: proliferative phase.   30X

1. Endometrial stroma   3. Lamina basalis
2. Lamina functionalis   (endometrium)
   (endometrium)   4. Uterine glands
                            5. Myometrium

Figure 17. 25 Uterus: secretory phase.   30X

1. Endometrium   4. Uterine glands (now
2. Myometrium   saw-toothed)
3. Endometrial stroma

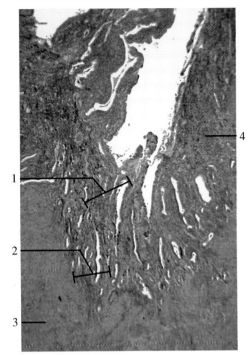

Figure 17. 26 Uterus: menstrual   30X
phase.
1. Disintegrating lamina   3. Myometrium
   functionalis   4. Blood in stroma
2. Lamina basalis (still
   intact)

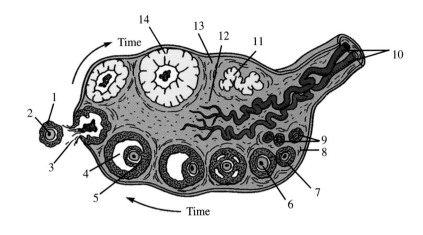

Figure 17.27 Structure of the ovary.
1. Corona radiata   8. Germinal epithelium
2. Secondary oocyte   9. Primary follicles
3. Ovulation   10. Ovarian vessels
4. Follicular fluid within antrum 11. Corpus albicans
5. Cumulus oophorus   12. Ovarian medulla
6. Oocyte   13. Ovarian cortex
7. Follicular cells   14. Corpus luteum

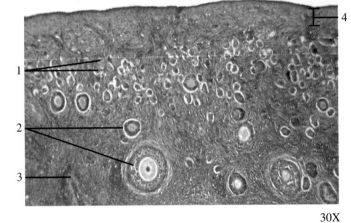

30X

Figure 17.28 Ovary.
1. Primordial follicles
2. Primary follicles
3. Atretic follicle
4. Tunical albuginea

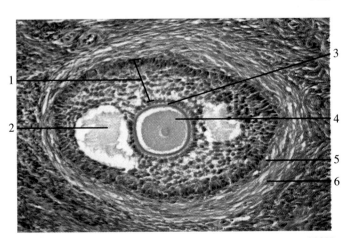

200X

Figure 17.29 Secondary follicle.
1. Granulosa cells
2. Antrum
3. Zona pellucida
4. Oocyte
5. Theca interna
6. Theca externa

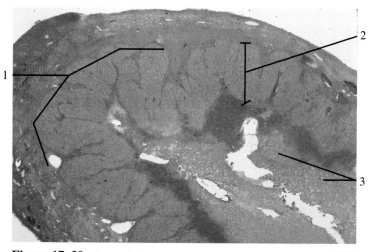

Figure 17. 30 Corpus luteum.   100X
1. Corpus luteum
2. Wall of corpus luteum
3. Former follicular antrum

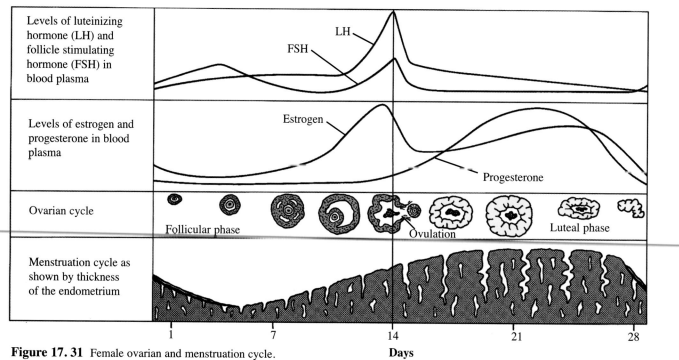

Figure 17. 31 Female ovarian and menstruation cycle.

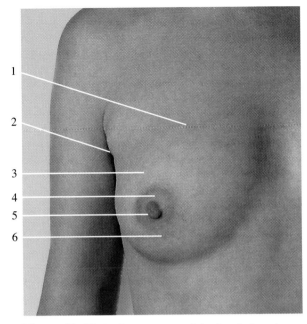

Figure 17. 32 Surface anatomy of the female breast.

1. Pectoralis major muscle
2. Axilla
3. Lateral process of breast
4. Areola
5. Nipple
6. Breast (containing mammary glands)

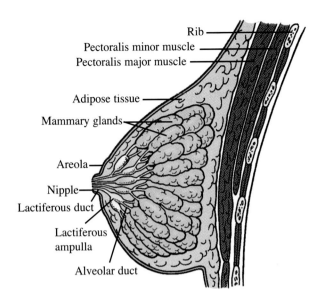

Figure 17. 33 Mammary gland.

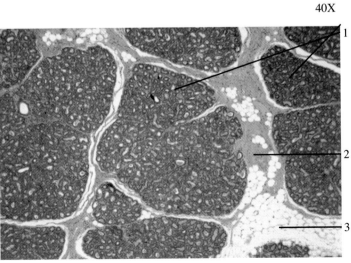

40X

Figure 17. 34 Mammary glands, (non-lactating glands).

1. Interlobular duct
2. Interlobular connective tissue
3. Lobule of glandular tissue
4. Intralobular connective tissue

400X

Figure 17. 35 Mammary glands, (lactating glands).

1. Lobules of glandular tissue
2. Intralobular connective tissue tissue
3. Adipose cells

Developmental Biology

18

The period of human pregnancy, which generally requires 38 weeks, is known as *gestation*. *Morphogenesis* is the sequence of changes that occur in the formation of the baby's body structures. Although gestation is frequently discussed chronologically as trimesters, prenatal development is more accurately divided morphogenically into three periods based on structural changes. On this basis, the *pre-embryonic period* includes the first two weeks following fertilization, the *embryonic period* includes the following six weeks, and the *fetal period* includes the final 30 weeks.

The events of the two-week pre-embryonic period include transportation of the fertilized egg, or *zygote*, through the uterine tube, mitotic divisions, implantation, and the formation of primordial embryonic tissue (fig. 18.1). Implantation begins between the fifth and seventh day and is made possible by the secretion of enzymes that digest a portion of the endometrium of the uterus. During implantation, the *trophoblast cells* secrete *human chorionic gonadotrophin* (hCG), which prevents the breakdown of the endometrium and menstruation. The trophoblast cells also participate in the formation of the placenta.

The events of the six-week embryonic period include the differentiation of the germ layers into specific body organs and the formation of the extraembryonic membranes, including the *placenta, umbilical cord, amnion, yolk sac, allantois,* and *chorion.*

A small amount of tissue differentiation and organ development occurs during the fetal period, but for the most part fetal development is primarily limited to body growth. Labor and parturition (childbirth) are the culmination of gestation and require the action of *oxytocin* from the mother's pituitary gland, and *prostaglandins*, produced in her uterus.

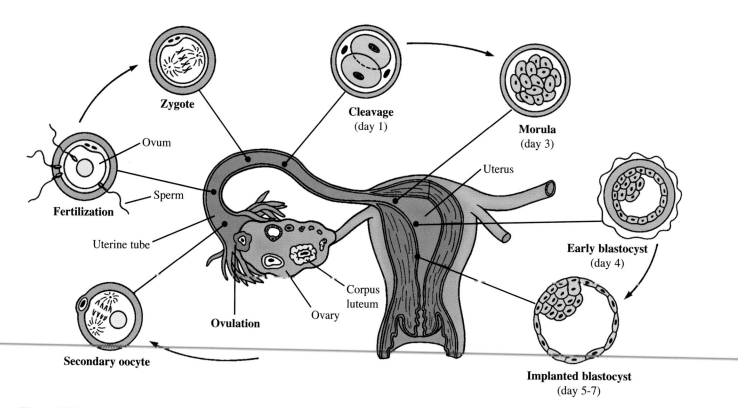

Figure 18.1 Events of ovulation, fertilization, and implantation.

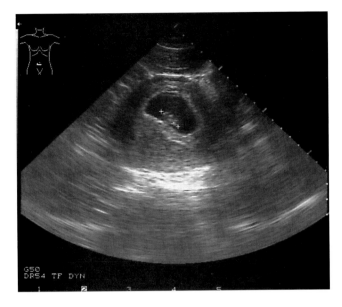

Figure 18.2 A 8.1 week intrauterine pregnancy. Marks are indicated on the image to denote the crown and rump. The crown-rump length of this embryo is 18 mm. Ultrasonography, produced by a mechanical vibration of high frequency, produces a safe, high resolution of fetal structure. Most ultrasound scans are obtained on fetuses older than 12 weeks.

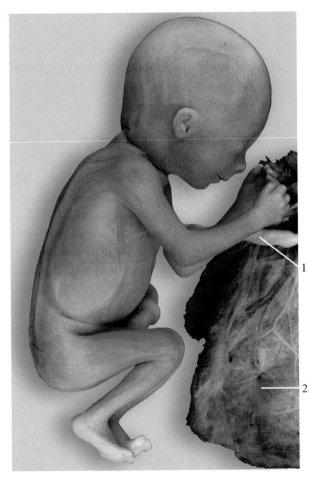

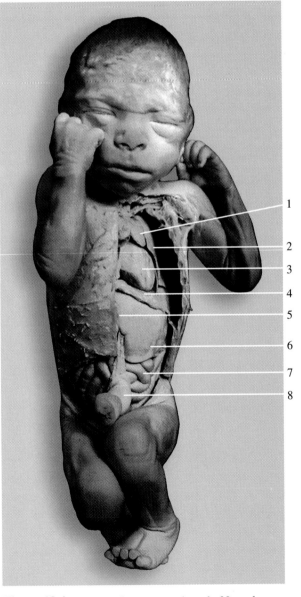

Figure 18.3 A human fetus at aproximately 24 weeks.
1. Umbilical cord
2. Placenta

Figure 18.4 A human fetus at aproximately 28 weeks.
1. Thymus   5. Falciform ligament
2. Lung   6. Liver
3. Heart   7. Small intestine
4. Diaphragm   8. Umbilical cord

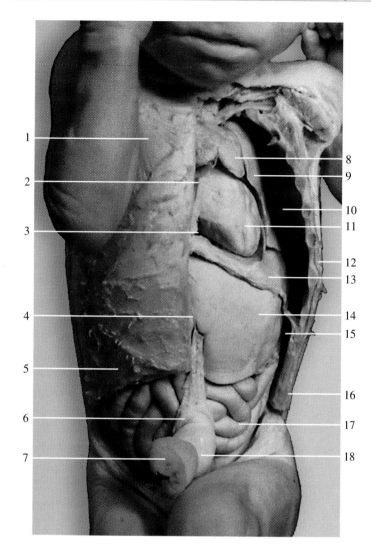

Figure 18.5 Thoracic and abdominal viscera of a human fetus at approximately 28 weeks.

1. Pectoralis major m.
2. Pericardium (cut)
3. Pericardial cavity
4. Falciform ligament
5. External abdominal oblique m.
6. Umbilicus
7. Umbilical vein
8. Thymus
9. Lung
10. Pleural cavity
11. Heart
12. Thoracic wall
13. Diaphragm
14. Liver
15. Peritoneal cavity
16. Abdominal wall
17. Small intestine
18. Umbilical cord

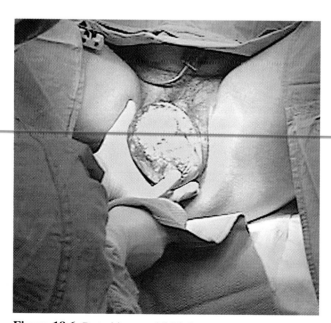

Figure 18.6 Parturition, or childbirth.

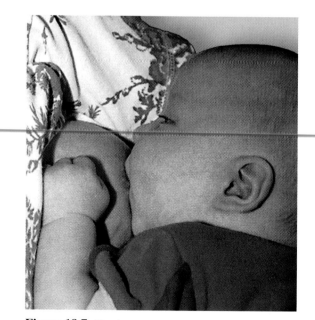

Figure 18.7 Nursing.

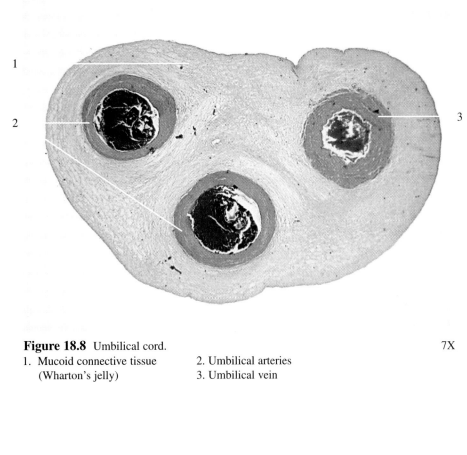

Figure 18.8 Umbilical cord.   7X

1. Mucoid connective tissue   2. Umbilical arteries
   (Wharton's jelly)   3. Umbilical vein

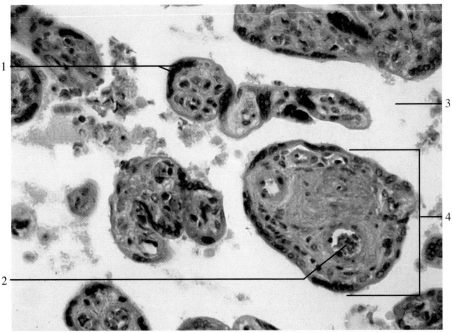

Figure 18.9 Placenta.   300X

1. Nuclei of syncytiotrophoblast   3. Intervillus space containing maternal
2. Blood vessel containing fetal blood cells   blood cells
                                          4. Choronic villus

Glossary of Prefixes and Suffixes

Element	Definition and Example	Element	Definition and Example	Element	Definition and Example
a-	absent, deficient or without: atrophy	circum-	around: circumduct	gastro-	stomach: gastrointestinal
ab-	off, away from: abduct	-cis	cut, kill: excision	-gen	an agent that produces or originates: pathogen
abdomin-	abdomen	co-	together: copulation		
-able	capable of: viable	coel-	hollow cavity: coelom	-genic	produced from, producing: carcinogenic
ac-	toward, to: actin	con-	with, together: congenital	gloss-	tongue: glossopharyngeal
acou-	hear, acoustic	contra-	against, opposite: contraception	glyco-	sugar: glycosuria
ad-	denoting to, toward: adduct			-gram	a record, recording: myogram
af-	movement toward a central point: afferent artery	corn-	denoting hardness: cornified	gran-	grain, particle: agranulocyte
		corp-	body: corpus	-graph	instrument for recording: electrocardiograph
alba-	pale or white: linea alba	crypt-	hidden: cryptorchism		
-alg	pain: neuralgia	cyan-	blue color: cyanosis	grav-	heavy: gravid
ambi-	both: ambidextrous	cysti-	sac or bladder: cystoscope	gyn-	female sex: gynocology
angi-	pertaining to vessel: angiology	cyto-	cell: cytology		
ante-	before: antebrachium	de-	down, from: descent	hema(o)-	blood: hematology
anti-	against: anticoagulant	derm-	skin: dermatology	haplo-	simple or single: haploid
aqua-	water: aqueous	di-	two: diarthrotic	hemi-	half: hemiplegia
archi-	to be first: archeteron	dipl-	double: diploid	hepat-	liver: hepatic portal
arthri-	joint: arthritis	dis-	apart, away from: disarticulate	hetero-	other, different: heterosexual
-asis	condition or state of: homeostasis	duct-	lead, conduct: ductus deferens	histo-	webb, tissue: histology
		dur-	hard: dura mater	holo-	whole, entire: holocrine
aud-	pertaining to ear: auditory	-dynia	pain	homo-	same, alike: homologous
auto-	self: autolysis	dys-	bad, difficult, painful: dysentery	hydro-	water: hydrocoel
				hyper-	beyond, above, excessive: hypertension
bi-	two: bipedal	e-	out, from: eccrine	hypo-	under, below: hypoglycemia
bio-	life: biology	ecto-	outside, outer, external: ectoderm		
blast-	generative or germ bud: osteoblast	-ectomy	surgical removal: tonsillectomy	-ia	state or condition: hypoglycemia
brachi-	arm: brachialis	ede-	swelling: edema	-iatrics	medical specialties: pediatrics
brachy-	short: brachydont	-emia	pertaining to a condition of the blood: lipemia	idio-	self, separate, distinct: idiopathic
brady-	slow: bradycardia			ilio-	ilium: iliosacral
bucc-	cheek: buccal cavity	end-	within: endoderm	infra-	beneath: infraspinatus
		entero-	intestine: enteritis	inter-	among, between
cac-	bad, ill: cachexia	epi-	upon, in addition: epidermis	intra-	inside, within: intracellular
calci-	stone: calculus	erythro-	red: erythrocyte	-ion	process: acromion
capit-	head: capitis	ex-	out of: excise	iso-	equal, like: isotonic
carcin-	cancer: carcinogenic	exo-	outside: exocrine	-ism	condition or state: rheumatism
cardi-	heart: cardiac	extra-	outside of, beyond, in addition: extracellular	-itis	inflammation: meningitis
caud-	tail: cauda equina			labi-	lip: labium majus
cata-	lower, under, against: catabolism	fasci-	band: fascia	lacri-	tears: nasolacrimal
-coel	swelling, and enlarged space or cavity: blastocoele	febr-	fever: febrile	later-	side: lateral
		-ferent	bear, carry: efferent arteriole	-logy	science of: morphology
cephal-	head: cephalis	fiss-	split: fissure	-lysis	solution, dissolve: hemolysis
cerebro-	brain: cerebrospinal fluid	for-	opening: foramen		
chol-	bile: cholic	-form	shape: fusiform	macro-	large, great: macrophage
chondr-	cartilage: chondrocyte			mal-	bad, abnormal, disorder: malignant
chrom-	color: chromocyte				
-cid	destroy: germicide				

Element	Definition and Example	Element	Definition and Example	Element	Definition and Example
medi-	middle: medial	-pnea	to breathe: apnea	trans-	across, over: transfuse
mega-	great, large: mega karyocyte	pneumato-	breathing: pneumonia	tri-	three: trigone
meso-	middle or moderate: mesoderm	pod-	foot: podiatry	trich-	hair: trichology
meta-	after, beyond: metatarsal	-poieis	formation of: hematopoiesis	-trophy	a state relating to nutrition: hypertrophy
micro-	small: microtome	poly-	many, much: polyploid	-tropic	turning toward, changing: gonadotropic
mito-	thread: mitosis	post-	after, behind: post natal		
mono-	alone, one, single: monocyte	pre-	before in time or place: prenatal	ultra-	beyond, excess: ultrasonic
mons-	mountain: mons pubis	prim-	first: primitive	uni-	one: unicellular
morph-	form, shape: morphology	pro-	before in time or place: prosect	uro-	urine, urinary organs or tract: uroscope
multi-	many, much: multinuclear	proct-	anus: proctology	-uria	urine: polyuria
myo-	muscle: myology	pseudo-	false: pseudostratified		
		psycho-	mental: psychology	vas-	vessel: vasoconstriction
narc-	numbness, stupor: narcotic	pyo-	pus: pyoculture	vermi-	worm: vermiform
neo-	new, young: neonatal			viscer-	organ: visceral
necro-	corpse, dead: necrosis	quad-	fourfold: quadriceps femoris	vit-	life: vitamin
nephro-	kidney: nephritis				
neuro-	nerve: neurolemma	re-	back, again: repolarization	zoo-	animal: zoology
noto-	back: notochord	rect-	straight: rectus abdominis	zygo-	union, join: zygote
		reno-	kidney: renal		
ob-	against, toward, in front of: obturator	rete-	network: retina		
oc-	against: occlusion	retro-	backward: retroperitoneal		
-oid	resembling, likeness: sigmoid	rhin-	nose: rhinitis		
oligo-	few, small: oligodendrocyte	-rrhage	excessive flow: hemorrhage		
-oma	tumor: lymphoma	-rrhea	flow, or discharge: diarrhea		
oo-	egg: oocyte				
or-	mouth: oral	sanguin-	blood: sanguiferous		
orchi-	testicles: cryptorchidism	sarc-	flesh: sarcoplasm		
-ory	pertaining to: sensory	-scope	instrument for examination of a part: stethoscope		
osteo-	bone: osteoblast	-sect	cut: dissect		
-ose	full of: adipose	semi-	half: semilunar		
oto	ear: otolith	serrate-	saw-edged: serratus anterior		
ovo-	egg: ovum	-sis	state or condition: dialysis		
		steno-	narrow: stenohaline		
para-	give birth to, bear: parturition	-stomy	surgical opening: tracheotomy		
para-	near, beyond, beside: paranasal	sub-	under, beneath, below: subcutaneous		
path-	disease, that which undergoes sickness: pathology	super-	above, beyond, upper: superficial		
-pathy	abnormality, disease: neuropathy	supra-	above, over: suprarenal		
ped-	children: pediatrician	syn (sym)	together, joined, with: synapse		
pen-	need, lack: penicillin				
-penia	deficiency: thrombocytopenia	tachy-	swift, rapid: tachometer		
per-	through: percutaneous	tele-	far: telencephalon		
peri-	near, around: pericardium	tens-	stretch: tensor fascia lata		
phag-	to eat: phagocyte	tetra-	four: tetrad		
-phil	have an affinity for: neutrophil	therm-	heat: thermogram		
phlebo-	vein: phlebitis	thorac-	chest: thoracic cavity		
-phobe	abnormal fear, dread: hydrophobia	thrombo-	lump, clot: thrombocyte		
-plasty	reconstruction of: rhinoplasty	-tomy	cut: appendectomy		
platy-	flat, side: platysma	tox-	poison: toxic		
-plegia	stroke, paralysis: paraplegia	tract-	draw, drag: traction		

Glossary

A

abdomen (ab-do´men): the portion of the trunk located between the diaphragm and the pelvis; contains the abdominal cavity and its visceral organs.

abduction (ab-duk´shun): a movement away from the axis or midline of the body; opposite of adduction; a movement of a digit away from the axis of a limb.

acapnia (ah-kap´ne-ah): a decrease in normal amount of CO_2 in the blood.

accommodation (ah-kom-o-da´shun): a change in the shape of the lens of the eye so that vision is more acute; the focusing for various distances.

acetone (as´e-tone): an organic compound that may be present in the urine of diabetics; also called ketone bodies.

Achilles tendon (ah-kil´ez): see *tendo calcaneus*.

acidosis (as-i-do´sis): a disorder of body chemistry in which the alkaline substances of the blood are reduced below normal.

actin (ak´tin): a protein in muscle fibers that together with myosin is responsible for contraction.

acoustic (ah-koos´tik): referring to sound or the sense of hearing.

adduction (ah-duk´shun): a movement toward the axis or midline of the body; opposite of abduction; a movement of a digit toward the axis of a limb.

adenohypophysis (ad´e-no-hi-pof´i-sis): anterior pituitary gland.

adenoid (ad´e-noid): paired lymphoid structures in the nasopharynx; also called pharyngeal tonsils.

adenosine triphosphate (ATP) (ah-den´o-sen tri-fos´fate): a chemical compound that provides energy for cellular use.

adipose (ad´e-pose): fat, or fat-containing, such as adipose tissue.

adrenal glands (ah-dre´nal): endocrine glands; one superior to each kidney; also called *suprarenal glands*.

aerobic (a-er-o´bik): requiring free O_2 for growth and metabolism as in the case of certain bacteria called *aerobes*.

allantois (ah-lan´to-is): an extraembryonic membranous sac that forms blood cells and gives rise to the fetal umbilical arteries and vein. It also contributes to the formation of the urinary bladder.

alveolus (al-ve´o-lus): an individual aircapsule within the lung. Alveoli are the basic functional units of respiration. Also, the socket that secures a tooth.

amnion (am´ne-on): a membrane that surrounds the fetus to contain the amniotic fluid.

amphiarthrosis (am´fe-ar-thro´sis): a slightly moveable joint in a functional classification of joints.

anatomical position (an´ah-tom´e-kal): an erect body stance with the eyes directed forward, the arms at the sides, and the palms of the hands facing forward.

anatomy (ah-nat´o-me): the branch of science concerned with the structure of the body and the relationship of its organs.

antebrachium (an´te-bra´ke-um): the forearm.

anterior (ventral) (an-te´re-or): toward the front; the opposite of *posterior (dorsal)*.

antigen (an´te-jen): a substance which causes cells to produce antibodies.

anus (a´nus): the terminal end of the GI tract, opening of the anal canal.

aorta (a-or´tah): the major systemic vessel of the arterial portion of the circulatory system, emerging from the left ventricle.

apocrine gland (ap´o-krin): a type of sweat gland that functions in evaporative cooling.

appendix (ah-pen´diks): a short pouch that attaches to the cecum.

aqueous humor (a´kwe-us hu´mor): the watery fluid that fills the anterior and posterior chambers of the eye.

arachnoid mater (ah-rak´noid): the weblike middle covering (meninx) of the central nervous system.

arbor vitae (ar´bor vi´tah): the branching arrangement of white matter within the cerebellum.

areola (ah-re´o-lah): the pigmented ring around the nipple.

artery (ar´ter-e): a blood vessel that carries blood away from the heart.

articular cartilage (ar-tik´u-lar ker´ti-lij): a hyaline cartilaginous covering over the articulating surface of bones of synovial joints.

ascending colon (ko´lon): the portion of the large intestine between the cecum and the right colic (hepatic) flexure.

atom (at´om): the smallest unit of an element that can exist and still have the properties of the element; collectively, atoms form molecules in a compound.

atrium (a´tre-um): either of two superior chambers of the heart that receive venous blood.

atrophy (at´ro-fe): a wasting away or decrease in size of a cell or organ.

auditory tube (aw´di-to´re): a narrow canal that connects the middle ear chamber to the pharynx; also called the *eustachian canal*.

autonomic (aw-to-nom´ik): self-governing; pertaining to the division of the nervous system which controls involuntary activities.

axilla (ak-sil´ah): the depressed hollow under the arm; the armpit.

axon (ak´son): The elongated process of a neuron (nerve cell) that transmits an impulse away from the cell body.

B

basement membrane: a thin sheet of extracellular substance to which the basal surfaces of membranous epithelial cells are attached.

basophil (ba´so-fil): a granular leukocyte that readily stains with basophilic dye.

belly: the thickest circumference of a skeletal muscle.

benign: (be-nine´): nonmalignant; a confined tumor.

blastula (blas´tu-lah): an early stage of prenatal development between the morula and embryonic stages.

blood: the fluid connective tissue that circulates through the cardiovascular system to transport substances throughout the body.

bolus (bo´lus): a moistened mass of food that is swallowed from the oral cavity into the pharynx.

bone: an organ composed of solid, rigid connective tissue, forming a component of the skeletal system.

Bowman's capsule (bo´manz kap´sul): see *glomerular capsule*.

brain: the enlarged superior portion of the central nervous system, located in the cranial cavity of the skull.

brain stem: the portion of the brain consisting of the medulla oblongata, pons, and midbrain.

bronchial tree (brong´ke-al): the bronchi and their branching bronchioles.

bronchiole (brong´ke-ol): a small division of a bronchus within the lung.

bronchus (bron´kus): a branch of the trachea that leads to a lung.

buccal cavity (buk´al): the mouth, or oral cavity.

bursa (ber´sah): a saclike structure filled with synovial fluid, which occurs around joints.

buttock (but´ok):the rump or fleshy mass on the posterior aspect of the lower trunk, formed primarily by the gluteal mucles.

C

calorie (kal´o-re): the unit of heat required to raise the temperature of one gram of water one degree centigrade.

calyx (ka´liks): a cup-shaped portion of the renal pelvis that encircles a renal papilla.

cancellous bone (kan´se-lus): spongy bone; bone tissue with a latticelike structure.

capillary (kap´i-lar´e): a microscopic blood vessel that connects an arteriole and a venule; the functional unit of the circulatory system.

carcinogenic (kar-si-no-jen´ik): stimulating or causing the growth of a malignant tumor, or cancer.

carpus (kar´pus): the proximal portion of the hand that contains the eight carpal bones.

cartilage (kar´ti´lij): a type of connective tissue with a solid elastic matrix.

caudal (kaw´dal): referring to a position more toward the tail.

cecum (se´kum): the pouchlike portion of the large intestine to which the ileum of the small intestine is attached.

cell: the structural and functional unit of an organism; the smallest structure capable of performing all the functions necessary for life.

central nervous system (CNS): the brain and the spinal cord.

centrosome (sen´tro-som): a dense body near the nucleus of a cell that contains a pair of centrioles.

cerebellum (ser´e-bel´um): the portion of the brain concerned with the coordination of movements and equilibrium.

cerebrospinal fluid (ser´e-bro-spi´nal): a fluid that buoys and cushions the central nervous system.

cerebrum (ser´e-brum): the largest portion of the brain, composed of the right and left hemispheres.

cervical (ser´vi-kal): pertaining to the neck or a necklike portion of an organ.

choanae (ko-a´na): the two posterior openings from the nasal cavity into the nasopharynx.

cholesterol (ko-les´ter-ol): an organic fat-like compound found in animal fat, bile, blood, liver, and other parts of the body.

chondrocyte (kon´dro-site): a cartilage cell.

chorion (ko´re-on): An extraembryonic membrane that participates in the formation of the placenta.

choroid (ko´roid): the vascular, pigmented middle layer of the wall of the eye.

chromosome (kro´mo-som): structure in the nucleus that contains the genes for genetic expression.

chyme (kime): The mass of partially digested food that passes from the stomach into the duodenum of the small intestine.

cilia (sil´e-ah): microscopic, hairlike processes that move in a wavelike manner on the exposed surfaces of certain epithelial cells.

ciliary body (sil´e-er´e): a portion of the choroid layer of the eye that secretes aqueous humor and contains the ciliary muscle.

circumduction (ser´kum-duk´shun): a conelike movement of a body part, such that the distal end moves in a circle while the proximal portion remains relatively stable.

clitoris (kli´to-ris): a small, erectile structure in the vulva of the female.

cochlea (kok´le-ah): the spiral portion of the inner ear that contains the spiral organ (organ of Corti).

ceolom (se´lom): the abdominal cavity.

colon (ko´lon): the first portion of the large intestine.

common bile duct: a tube that is formed by the union of the hepatic duct and cystic duct, transports bile to the duodenum.

compact bone: tightly packed bone that is superficial to spongy bone; also called *dense bone*.

condyle (kon´dile): a rounded process at the end of a long bone that forms an articulation.

connective tissue: one of the four basic tissue types within the body. It is a binding and supportive tissue with abundant matrix.

cornea (kor´ne-ah): the transparent convex, anterior portion of the outer layer of the eye.

cortex (kor´teks): the outer layer of an organ such as the convoluted cerebrum, adrenal gland, or kidney.

costal cartilage (kos´tal): the cartilage that connects the ribs to the sternum.

cranial (kra′ne-al): pertaining to the cranium.

cranial nerve: one of twelve pairs of nerves that arise from the inferior surface of the brain.

D

dentin (den′tine): the principal substance of a tooth, covered by enamel over the crown and by cementum on the root.

dermis (der′mis): the second, or deep, layer of skin beneath the epidermis.

descending colon: the segment of the large intestine that descends on the left side from the level of the spleen to the level of the left iliac crest.

diaphragm (di′ah-fram): a flat dome of muscle and connective tissue that separates the thoracic and abdominal cavities.

diaphysis (di-af′i-sis): the shaft of a long bone.

diastole (di-as′to-le): the sequence of the cardiac cycle during which the ventricular heart chamber wall is relaxed.

diarthrosis (di′ar-thro′sis): a freely movable joint.

distal (dis′tal): away from the midline or origin; the opposite of *proximal*.

dorsal (dor′sal): pertaining to the back or posterior portion of a body part; the opposite of *ventral*.

ductus deferens (duk′tus def′er-enz): a tube that carries spermatozoa from the epididymis to the ejaculatory duct: also called the *vas deferens* or *seminal duct*.

duodenum (du′o-num): the first portion of the small intestine.

dura mater (du′rah ma′ter): the outermost meninx covering the central nervous system.

E

eccrine gland (ek′rin): a sweat gland that functions in body cooling.

ectoderm (ek′to-derm): the outermost of the three primary embryonic germ layers.

edema (e-de′mah): an excessive retention of fluid in the body tissues.

effector (ef-fek′tor): an organ such as a gland or muscle that responds to motor stimulation.

efferent (ef′er-ent): conveying away from the center of an organ or structure.

ejaculation (e-jak′u-la′shun): the discharge of semen from the male urethra during climax.

electrocardiogram (e-lek′tro-kar′de-o-gram′): a recording of the electrical activity that accompanies the cardiac cycle; also called ECG or EKG.

electroencephalogram (e-lek′tro-en-sef′ah-lo-gram): a recording of the brain wave pattern; also called EEG.

electromyogram (e-lek′tro-mi′o-gram): a recording of the activity of a muscle during contraction: also called EMG.

electrolyte (e-lek′tro-lite): a solution that conducts electricity by means of charged ions.

electron (e-lek′tron): the unit of negative electricity.

enamel (en-am′el): the outer, dense substance covering the crown of a tooth.

endocardium (en′do-kar′de-um): the fibrous lining of the heart chambers and valves.

endochondral bone (en′do-kon′dral): bones that form as hyaline cartilage models first and then are ossified.

endocrine gland (en′do-krine): hormone producing gland that is part of the endocrine system.

endoderm (en′do-derm): the innermost of the three primary germ layers of an embryo.

endometrium (en′do-me′tre-um): the inner lining of the uterus.

endothelium (en′do-the′le-um): the layer of epithelial tissue that forms the thin inner lining of blood vessels and heart chambers.

eosinophil (e′o-sin′o-fil): a type of white blood cell that becomes stained by acidic eosin dye; constitutes about 2%–4% of the white blood cells.

epicardium (ep′i-kar′de-um): the thin, outer layer of the heart: also called the *visceral pericardium*.

epidermis (ep′i-der′mis): the outermost layer of the skin, composed of stratified squamous epithelium.

epididymis (ep′i-did′i-mis): a coiled tube located along the posterior border of the testis; stores spermatozoa and discharges them during ejaculation.

epidural space (ep′i-du′ral): a space between the spinal dura mater and the bone of the vertebral canal.

epiglottis (ep′i-glot′is): a leaflike structure positioned on top of the larynx that covers the glottis during swallowing.

epinephrine (ep′i-nef′rin): a hormone secreted from the adrenal medulla resulting in action similar to those from sympathetic nervous system stimulation; also called *adrenaline*.

epiphyseal plate (ep′i-fize-al): a cartilaginous layer located between the epiphysis and diaphysis of a long bone and functions in　　longitudinal bone growth.

epiphysis (e-pif′i-sis): the end segment of a long bone, distinct in early life but later becoming part of the larger bone.

epithelial tissue (ep′i-the′le-al): one of the four basic tissue types; the type of tissue that covers or lines all exposed body surfaces.

erythrocyte (e-rith′ro-site): a red blood cell.

esophagus (e-sof′ah-gus): a tubular organ of the GI tract that leads from the pharynx to the stomach.

estrogen (es′tro-jen): female sex hormone secreted from the ovarian (Graafian) follicle.

eustachian canal (u-sta′ke-an): see *auditory tube*.

excretion (eks-kre′shun): discharging waste material.

exocrine gland (ek′so-krin): a gland that secretes its product to an epithelial surface, directly or through ducts.

expiration (ek′spi-ra′shun): the process of expelling air from the lungs through breathing out; also called *exhalation*.

extension (ek-sten´shun): a movement that increases the angle between two bones of a joint.

external ear: the outer portion of the ear, consisting of the auricle (pinna), external auditory canal.

extracellular (esk-trah-sel´u-lar): outside a cell or cells.

extrinsic (eks-trin´sik): pertaining to an outside or external origin.

F

facet (fas´et): a small, smooth surface of a bone where articulation occurs.

fallopian tube (fal-lo´pe-an): see *uterine tube*.

fascia (fash´e-ah): a tough sheet of fibrous connective tissue binding the skin to underlying muscles or supporting and separating muscle.

fasciculus (fah-sik´u-lus): a bundle of muscle or nerve fibers.

feces (fe´sez): waste material expelled from the GI tract during defecation, composed of food residue, bacteria, and secretions; also called *stool*.

fetus (fe´tus): the unborn offspring during the last stage of prenatal development.

filtration (fil-tra´shun): the passage of a liquid through a filter or a membrane.

fimbriae (fim´bre-e): fringelike extensions from the borders of the open end of the uterine tube.

fissure (fish´ure): a groove or narrow cleft that separates two parts of an organ.

flexion (flek´shun): a movement that decreases the angle between two bones of a joint; opposite of extension.

fontanel (fon´tah-nel): a membranous-covered region on the skull of a fetus or baby where ossification has not yet occurred: also called a *soft spot*.

foot: the terminal portion of the lower extremity, consisting of the tarsus, metatarsus, and digits.

foramen (fo-ra´men): an opening in an anatomical structure for the passage of a blood vessel or a nerve.

foramen ovale (o-val´e): the opening through the interatrial septum of the fetal heart.

fossa (fos´ah): a depressed area, usually on a bone.

fourth ventricle (ven´tri-k´l): a cavity within the brain containing cerebrospinal fluid.

fovea centralis (fo´ve-ah sen´tra´lis): a depression on the macula lutea of the eye where only cones are located, which is the area of keenest vision.

G

gallbladder: a pouchlike organ, attached to the inferior side of the liver, which stores and concentrates bile.

gamete (gam´ete): a haploid sex cell, sperm or egg.

gamma globulins (gam´mah glob´u-lins): protein substances often found in immune serums that act as antibodies.

ganglion (gang´gle-on): an aggregation of nerve cell bodies outside the central nervous system.

gastrointestinal tract (gas´tro-in-tes´tin-al): the tubular portion of the digestive system that includes the stomach and the small and large intestines; also called the *GI tract*.

gene (jene): one of the biologic units of heredity; parts of the DNA molecule located in a definite position on a certain chromosome.

genetics (je-net´iks): the study of heredity.

gingiva (jin-ji´vah): the fleshy covering over the mandible and maxilla through which the teeth protrude within the mouth; also called the *gum*.

gland: an organ that produces a specific substance or secretion.

glans penis (glanz pe´nis): the enlarged, distal end of the penis.

glomerular capsule (glo-mer´u-lar): the double-walled proximal portion of a renal tubule that encloses the glomerulus of a *nephron;* also called *Bowman's capsule*.

glomerulus (glo-mer´u´lus): a coiled tuft of capillaries that is surrounded by the glomerular capsule and filters urine from the blood.

glottis (glot´is): a slitlike opening into the larynx, positioned between the vocal folds.

glycogen (gli´ko-jen): the principal storage carbohydrate in animals. It is stored primarily in the liver and is made available as glucose when needed by the body cells.

goblet cell: a unicellular gland within columnar epithelia that secretes mucus.

gonad (go´nad): a reproductive organ, testis or ovary, that produces gametes and sex hormones.

gray matter: the portion of the central nervous system that is composed of nonmyelinated nervous tissue.

greater omentum (o-men´tum): a double-layered peritoneal membrane that originates on the greater curvature of the stomach and extends over the abdominal viscera.

gut: pertaining to the intestine; generally a developmental term.

gyrus (ji´rus): a convoluted elevation or ridge.

H

hair: an epidermal structure consisting of keratinized dead cells that have been pushed up from a dividing basal layer.

hair cells: specialized receptor nerve endings for responding to sensations, such as in the spiral organ of the inner ear.

hair follicle (fol´li-k´l): a tubular depression in the skin in which a hair develops.

hand: the terminal portion of the upper extremity, consisting of the carpus, metacarpus, and digits.

hard palate (pal´at): the bony partition between the oral and nasal cavities, formed by the maxillae and palatine bones.

haustra (haws´trh): sacculations or pouches of the colon.

haversian system (ha-ver'shan): see *osteon*.

heart: a muscular, pumping organ positioned in the thoracic cavity

hematocrit (he-mat'o-krit): the volume percentage of red blood cells in whole blood.

hemoglobin (he'mo-glo'bin): the pigment of red blood cells that transports O_2 and CO_2.

hemopoiesis (hem'ah-poi-e'sis): production of red blood cells.

hepatic portal circulation (por'tal): the return of venous blood from the digestive organs and spleen through a capillary network within the liver before draining into the heart.

heredity (re-red'i-te): the transmission of certain characteristics, or traits, from parents to offspring, via the genes.

hiatus (hi-a'tus): an opening or fissure.

hilum (hi'lum): a concave or depressed area where vessels or nerves enter or exit an organ.

histology (his-tol'o-je): microscopic anatomy of the structure and function of tissues.

homeostasis (ho-me-o-sta'sis): a consistency and uniformity of the internal body environment which maintains normal body function.

hormone (hor'mone): a chemical substance that is produced in an endocrine gland and secreted into the bloodstream to cause an effect in a specific target organ.

hyaline cartilage (hi'ah-line): the most common kind of cartilage in the body, occurring at the articular ends of bones, in the trachea, and within the nose, and forms the precursor to most of the bones of the skeleton.

hymen (hi'men): a developmental remnant (vestige) of membranous tissue that partially covers the vaginal opening.

hyperextension (hi'per-ek-sten'shun): extension beyond the normal anatomical position of 180°.

hypothalamus (hi'po-thal'ah-mus): a structure within the brain below the thalamus, which functions as an autonomic center and regulates the pituitary gland.

I

ileocecal valve (il'e-o-se'kal): a specialization of the mucosa at the junction of the small and large intestine that forms a one-way passage and prevents the backflow of food materials.

ileum (il'e-um): the terminal portion of the small intestine between the jejunum and cecum.

inferior vena cava (ve'nah ka'vah): a systemic vein that collects blood from the body regions inferior to the level of the heart and returns it to the right atrium.

inguinal (ing'gwi-nal): pertaining to the groin region.

inguinal canal: the passage in the abdominal wall through which a testis descends into the scrotum.

insertion: the more movable attachment of a muscle, usually more distal in location.

inspiration (in'spi-ra'shun): the act of breathing air into the alveoli of the lungs; also called *inhalation*.

integument (in-teg'u-ment): pertaining to the skin.

internal ear: the innermost portion or chamber of the ear, containing the cochlea and the vestibular organs.

interstitial (in-ter-stish'al): pertaining to spaces or structures between the functioning active tissue of any organ.

intracellular (in-trah-sel'u-lar): within the cell itself.

intervertebral disc (in'ter-ver'te-bral): a pad of fibrocartilage between the bodies of adjacent vertebrae.

intestinal gland (in-tes'ti-nal): a simple tubular digestive gland that opens onto the surface of the intestinal mucosa and secretes digestive enzymes; also called *crypt of Lieberkuhn*.

intrinsic (in-trin'sik): situated in or pertaining to an internal organ.

iris (i'ris): the pigmented, vascular tunic portion of the eye that surounds the pupil and regulates its diameter.

islets of Langerhans (i'lets of lahng'er-hanz): see *pancreatic islets*.

isotope (i'so-tope): a chemical element that has the same atomic number as another but a different atomic weight.

isthmus (is'mus): a narrow neck or portion of tissue connecting two structures.

J

jejunum (je-joo'num): the middle portion of the small intestine, located between the duodenum and the ileum.

joint capsule (kap'sule): the fibrous tissue that encloses the joint cavity of a synovial joint.

jugular (jug'u-lar): pertaining to the veins of the neck which drain the areas supplied by the carotid arteries.

K

karyotype (kar'e-o-tip): the arrangement of chromosomes that is characteristic of the species or of a certain individual.

keratin (ker'ah-tin): an insoluble protein present in the epidermis and in epidermal derivatives such as hair and nails.

kidney (kid'ne): one of the paired organs of the urinary system that contains nephrons and filters wastes from the blood in the formation of urine.

L

labia majora (la'be-ah ma-jor'ah): a portion of the external genitalia of a female, consisting of two longitudinal folds of skin extending downward and backward from the mons pubis.

labia minora (ma-nor'ah): two small folds of skin, devoid of hair and sweat glands, lying between the labia majora of the external genitalia of a female.

lacrimal gland (lak'ri-mal): a tear-secreting gland, located on the superior lateral portion of the eyeball underneath the upper eyelid.

lactation (lak-ta'shun): the production and secretion of milk by the mammary glands.

lacteal (lak´te-al): a small lymphatic duct within a villus of the small intestine.

lacuna (lah-ku´nah): a hollow chamber that houses an osteocyte in mature bone tissue or a chondrocyte in cartilage tissue.

lamella (lah-mel´ah): a concentric ring of matrix surrounding the central canal in an osteon of mature bone tissue.

large intestine: the last major portion of the GI tract, consisting of the cecum, colon, rectum, and anal canal.

larynx (lar´inks): the structure located between the pharynx and trachea that houses the vocal folds (cords); commonly called the *voice box*.

lens (lenz): a transparent refractive structure of the eye, derived from ectoderm and positioned posterior to the pupil and iris.

leukocyte (lu´ko-site): a white blood cell; also spelled *leucocyte*.

ligament (lig´ah-ment): a fibrous band or cord of connective tissue that binds bone to bone to strengthen and provide support to the joint; also may support viscera.

limbic system (lim´bik): a portion of the brain concerned with emotions and autonomic activity.

linea alba (lin´e-ah al´bah): a fibrous band extending down the anterior medial portion of the abdominal wall.

locus (lo´kus): the specific location or site of a gene within the chromosome.

lumbar (lum´bar): pertaining to the region of the loins.

lumen (lu´men): the space within a tubular structure through which a substance passes.

lung: one of the two major organs of respiration within the thoracic cavity.

lymph (limf): a clear fluid that flows through lymphatic vessels.

lymph node: a small, ovoid mass located along the course of lymph vessels.

lymphocyte (lim´fo-site): a type of white blood cell characterized by a granular cytoplasm.

M

macula lutea (mak´u-lah lu´te-ah): a depression in the retina that contains the fovea centralis, the area of keenest vision.

malignant (mah-lig´nant): a disorder that becomes worse and eventually causes death, as in cancer.

malnutrition (mal-nu-trish´un): any abnormal assimilation of food; receiving insufficient nutrients.

mammary gland (mam´er-e): the gland of the female breast responsible for lactation and nourishment of the young.

marrow (mar´o): the soft vascular tissue that occupies the inner cavity of certain bones and produces blood cells.

matrix (ma´triks): the intercellular substance of a tissue.

meatus (me-a´tus): an opening or passageway into a structure.

mediastinum (me´de-as-ti´num): the partition in the center of the thorax between the two pleural cavities.

medulla (me-dul´ah): the center portion of an organ.

medulla oblongata (ob´long-ga´tah): a portion of the brain stem between the pons and the spinal cord.

medullary cavity (med´u-lar´e): the hollow center of the diaphysis of a long bone, occupied by marrow.

meiosis (mi-o´sis): cell division by which gametes, or haploid sex cells, are formed.

melanocyte (mel´ah-no-site): a pigment-producing cell in the deepest epidermal layer of the skin.

membranous bone (mem´brah-nus): bone that forms from membranous connective tissue rather than from cartilage.

menarche (me-nar´ke): the first menstrual discharge.

meninges (me-nin´jez): a group of three fibrous membranes that cover the central nervous system.

menisci (men-is´si): wedge-shaped cartilages in certain synovial joints.

menopause (men´o-pawz): the cessation of menstrual periods in the human female.

menses (men´sez): the monthly flow of blood from the female genital tract.

menstrual cycle (men´stru-al): the rhythmic female reproductive cycle, characterized by changes in hormone levels and physical changes in the uterine lining.

menstruation (men´stru-a´shun): the discharge of blood and tissue from the uterus at the end of the menstrual cycle.

mesentery (mes´en-ter´e): a fold of peritoneal membrane that attaches an abdominal organ to the abdominal wall.

mesoderm (mes´o-derm): the middle one of the three primary germ layers.

mesothelium (mes´o-the´leum): a simple squamous epithelial tissue that lines body cavities and covers visceral organs; also called *serosa*.

metabolism (me-tab´o-lizm): the chemical changes that occur within a cell.

metacarpus (met´ah-kar´pus): the region of the hand between the wrist and the digits, including the five bones that support the palm of the hand.

metastasis (me-tas´tah-sis): the spread of a disease from one organ or body part to another.

metatarsus (met´ah-tar´sus): the region of the foot between the ankle and the digits, containing five bones.

microbiology (mi-kro-bi-ol´o-je): the science dealing with microscopic organisms, including bacteria, fungi, viruses, and protozoa.

microvilli (mi´kro-vil´i): microscopic, hairlike projections of cell membranes on certain epithelial cells.

midbrain: the portion of the brain between the pons and the forebrain.

middle ear: the middle of the three ear chambers, containing the three auditory ossicles.

mitosis (mi-to´sis): the process of cell division, in which the two daughter cells are identical and contain the same number of chromosomes.

mitral valve (mi´tral): the left atrioventricular heart valve; also called the bicuspid valve.

mixed nerve: a nerve containing both motor and sensory nerve fibers.

molecule (mol'e-kule): a minute mass of matter, composed of a combination of atoms that form a given chemical substance or compound.

motor neuron (nu´ron): a nerve cell that conducts action potential away from the central nervous system and innervates effector organs (muscles and glands); also called *efferent neuron*.

motor unit: a single motor neuron and the muscle fibers it innervates.

mucosa (mu-ko´sah): a mucous membrane that lines cavities and tracts opening to the exterior.

muscle (mus´el): an organ adapted to contract; three types of muscle tissue are cardiac, smooth, and skeletal.

mutation (mu-ta´shun): a variation in an inheritable characteristic, a permanent transmissible change in which the offspring differ from the parents.

myelin (me´e-lin): a lipoprotein material that forms a sheathlike covering around nerve fibers.

myocardium (mi´o-kar´de-um): the cardiac muscle layer of the heart.

myofibril (mi´o-fi´bril): a bundle of contractile fibers within muscle cells.

myoneural junction (mi´o-nu´ral): the site of contact between an axon of a motor neuron and a muscle fiber.

myosin (mi´o-sin): a thick filament protein that together with actin causes muscle contraction.

N

nail: a hardenend, keratinized plate that develops from the epidermis and forms a protective covering on the dorsal surfaces of the digits.

nares (na´rez): the opening into the nasal cavity; also called *nostrils*.

nasal cavity (na´zal): a mucosa-lined space above the oral cavity, which is divided by a nasal septum and is the first chamber of the respiratory system.

nasal septum (sep´tum): a bony and cartilaginous partition that separates the nasal cavity into two portions.

nephron (nef´ron): the functional unit of the kidney, consisting of a glomerulus, glomerular capsule, convoluted tubules, and the loop of the nephron.

nerve: a bundle of nerve fibers outside the central nervous system.

neurofibril node (nu´ro-fi´bril): a gap in the myelin sheath of a nerve fiber; also called the *node of Ranvier*.

neuroglia (nu-rog´le-ah): specialized supportive cells of the central nervous system.

neurolemmocyte (nu´ri-lem-o´site): a specialized neuroglia cell that surrounds an axon fiber of a peripheral neuron and forms the neurilemmal sheath; also called the *Schwann cell*.

neuron (nu´ron): the structural and functional unit of the nervous system, composed of a cell body, dendrites, and an axon; also called a *nerve cell*.

neutrophil (nu´tro-fil): a type of phagocytic white blood cell.

nipple: a dark pigmented, rounded projection at the tip of the breast.

node of Ranvier (rah-ve-a´): see *neurofibril node*.

notochord (no´to-kord): a flexible rod of tissue that extends the length of the back of an embryo.

nucleus (nu´kle-us): a spheroid body within a cell that contains the genetic factors of the cell.

nurse cells: specialized cells within the testes that supply nutrients to developing spermatozoa; also called *sertoli cells* or *sustentacular cells*.

O

olfactory (ol-fak´to-re): pertaining to the sense of smell.

oocyte (o´o-site): a developing egg cell.

oogenesis (o´o-hen´e-sis): the process of female gamete formation.

optic (op´tik): pertaining to the eye and the sense of vision.

optic chiasma (ki-as´mah): an X-shaped structure on the inferior aspect of the brain where there is a partial crossing over of fibers in the optic nerves.

optic disc: a small region of the retina where the fibers of the ganglion neurons exit from the eyeball to form the optic nerve; also called the *blind spot*.

oral: pertaining to the mouth; also called *buccal*.

organ: a structure consisting of two or more tissues, which performs a specific function.

organelle (or´gan-el´): a minute structure of a cell with a specific function.

organism (or´gah-nizm): an individual living creature.

orifice (or´i fis): an opening into a body cavity or tube.

origin (or´i-jin): the place of muscle attachment onto the more stationary point or proximal bone; opposite the insertion.

osmosis (os-mo´sis): the passage of a solvent, such as water, from a solution of lesser concentration to one of greater concentration through a semipermeable membrane.

ossicle (os´si-l´l): one of the three bones of the middle ear.

osteocyte (os´te-o-site): a mature bone cell.

osteon (os´te-on): a group of osteocytes and concentric lamellae surrounding a central canal within bone tissue; also called a *haversian system*.

ovarian follicle (o-va´re-an fol´li-k´l): a developing ovum and its surrounding epithelial cells.

ovary (o´vah-re): the female gonad in which ova and certain sexual hormones are produced.

oviduct (o´vi-dukt): the tube that transports ova from the ovary to the uterus; also called the *uterine tube* or *fallopian tube*.

ovulation (o´vu-la´shun): the rupture of an ovarian follicle with the release of an ovum.

ovum (o′vum): a secondary oocyte after ovulation but before fertilization.

P

palate (pal′at): the roof of the oral cavity.

palmar (pal′mar): pertaining to the palm of the hand.

pancreas (pah′kre-as): organ in the abdominal cavity that secretes gastric juices into the GI tract and insulin and glucagon into the blood.

pancreatic islets (pan′kre-at′ik): a cluster of cells within the pancreas that forms the endocrine portion of the pancreas; also called *islets of Langerhans*.

papillae (pah-pil′e): small nipplelike projections.

paranasal sinus (par′ah-na′zal si′nus): an air chamber lined with a mucous membrane that communicates with the nasal cavity.

parasympathetic (par′ah-smi′pah-thet′ik): pertaining to the division of the autonomic nervous system concerned with activities that are antagonistic to sympathetic.

parathyroids (par′ah-thi′roids): small endocrine glands that are embedded on the posterior surface of the thyroid glands and are concerned with calcium metabolism.

parietal (pah-ri′e-tal): pertaining to a wall of an organ or cavity.

parotid gland: (pah-rot′id): one of the paired salivary glands on the side of the face over the masseter muscle.

parturition (par′tu-rish′un): the process of childbirth.

pathogen (path′o-jen): any disease-producing organism.

pectoral girdle (pek′to-ral): the portion of the skeleton that supports the upper extremities.

pelvic (pel′vik): pertaining to the pelvis.

pelvic girdle: the portion of the skeleton to which the lower extremities are attached.

penis (pe′nis): the external male genital organ, through which urine passes during urination and which transports semen to the female during coitus.

pericardium (per′i-kar′de-um): a protective serous membrane that surrounds the heart.

perineum (per′i-ne′um): the floor of the pelvis, which is the region between the anus and the scrotum in the male and between the anus and the vulva in the female.

periosteum (per′e-os′te-um): a fibrous connective tissue covering the outer surface of bone.

peripheral nervous system (pe-rif′er-al): the nerves and ganglia of the nervous system that lie outside of the brain and spinal cord.

peristalsis (per′i-stal′sis): rhythmic contractions of smooth muscle in the walls of various tubular organs, which move the contents along.

peritoneum (per′i-to-ne′um): the serous membrane that lines the abdominal cavity and covers the abdominal viscera.

phagocyte (fag′o-site): any cell that engulfs other cells, including bacteria, or small foreign particles.

phalanx (fa′lanks), pl. *phalanges*: a bone of the finger or toe.

pharynx (far′inks): the organ of the GI tract and respiratory system located at the back of the oral and nasal cavities and extending to the larynx anteriorly and the esophagus posteriorly; also called the *throat*.

physiology (fiz′e-ol′o-je): the science that deals with the study of body functions.

pia mater (pi′ah ma′ter): the innermost meninx that is in direct contact with the brain and spinal cord.

pineal gland (pin′e-al): a small cone-shaped gland located in the roof of the third ventricle.

pituitary gland (pi-tu′i-tar′e): a small, pea-shaped endocrine gland situated on the inferior surface of the brain that secretes a number of hormones; also called the *hypophysis* and commonly called the "*master gland*."

placenta (plah-sen′tah): the organ of metabolic exchange between the mother and the fetus.

plasma (plaz′mah): the fluid, extracellular portion of circulating blood.

platelets (plate′lets): fragments of specific bone marrow cells that function in blood coagulation: also called *thrombocytes*.

pleural membranes (ploor′al): serous membranes that surround the lungs and line the thoracic cavity.

plexus (plek′sus): a network of enterlaced nerves or vessels.

plica circulares (pli′kah ser-ku-lar′is): a deep fold within the wall of the small intestine that increases the absorptive surface area.

pons (ponz): the portion of the brain stem just above the medulla oblongata and anterior to the cerebellum.

posterior (dorsal) (pos-ter′e-or): toward the back.

pregnancy: a condition where a female has a developing offspring in the uterus.

prenatal (pre-na′tal): the period of offspring development during pregnancy; before birth.

proprioceptor (pro′pre-o-sep′tor): a sensory nerve ending that responds to changes in tension in a muscle or tendon.

prostate (pros′tate): a walnut-shaped gland surrounding the male urethra just below the urinary bladder that secretes an additive to seminal fluid during ejaculation.

proximal (prok′si-mal): closer to the midline of the body or origin of an appendage: opposite of *distal*.

puberty (pu′ber-te): the period of development in which the reproductive organs become functional.

pulmonary (pul′mo-ner′e): pertaining to the lungs.

pupil: the opening through the iris that permits light to enter the vitreous chamber of the eyeball and be refracted by the lens.

R

receptor (re-sep′tor): a sense organ or a specialized end of a sensory neuron that receives stimuli from the environment.

rectum (rek′tum): the portion of the GI tract between the sigmoid colon and the anal canal.

reflex arc: the basic conduction pathway through the nervous system, consisting of a sensory neuron, interneuron, and a motor neuron.

renal (re´nal): pertaining to the kidney.

renal corpuscle (kor´pus´l): the portion of the nephron consisting of the glomerulus and a glomerular capsule.

renal pelvis: the inner cavity of the kidney formed by the expanded ureter and into which the calyces open.

replication (re-pli-ka´shun): the process of producing a duplicate; a copying or duplication, such as DNA replication.

respiration: (res´pi-ra´shun): the exchange of gases between the external environment and the cells of an organism.

rete testis (re´te tes´tis): a network of ducts in the center of the testis, site of spermatozoa production.

retina (ret´i-nah): the inner layer of the eye that contains the rods and cones.

retraction (re-trak´shun): the movement of a body part, such as the mandible, backward on a plane parallel with the ground; the opposite of *protraction*.

rod: a photoreceptor in the retina of the eye that is specialized for colorless, dim light vision.

rotation (ro-ta´shun): the movement of a bone around its own longitudinal axis.

rugae (ru´je): the folds or ridges of the mucosa of an organ.

S

sagittal (saj´i-tal): a vertical plane through the body that divides it into right and left portions.

salivary gland (sal´i-ver-e): an accessory digestive gland that secretes saliva into the oral cavity.

sarcolemma (sar´ko-lem´ah): the cell membrane of a muscle fiber.

sarcomere (sar´ko-mere): the portion of a skeletal muscle fiber between the two adjacent Z lines that is considered the functional unit of a myofibril.

Schwann cell (shwahn): see *neurolemmocyte*.

sclera (skle´rah): the outer white layer of connective tissue that forms the protective covering of the eye.

scrotum (skro´tum): a pouch of skin that contains the testes and their accessory organs.

sebaceous gland (se-ba´shus): an exocrine gland of the skin that secretes *sebum*, an oily protective product.

semen (se´men): the secretion of the reproductive organs of the male, consisting of spermatozoa and additives.

semicircular ducts: tubular channels within the inner ear that contain the receptors for equilibrium; also called *semicircular canals*.

semilunar valve (sem´e-lu´nar): crescent-shaped heart valves, positioned at the entrances to the aorta and the pulmonary trunk.

seminal vesicles (sem´i-nal ves´i-k´lz): a pair of accessory male reproductive organs lying posterior and inferior to the urinary bladder, which secrete additives to spermatozoa into the ejaculatory ducts.

sensory neuron (nu´ron): a nerve cell that conducts an impulse from a receptor organ to the central nervous system; also called *afferent neuron*.

serous membrane (se´rus): an epithelial and connective tissue membrane that lines body cavities and covers viscera; also called *serosa*.

sesamoid bone (ses´ah-moid): a membranous bone formed in a tendon in response to joint stress.

sigmoid colon (sig´moid ko´lon): the S-shaped portion of the large intestine between the descending colon and the rectum.

sinoatrial node (sin´o-a´tre-al): a mass of cardiac tissue in the wall of the right atrium that initiates the cardiac cycle; the SA node; also called the *pacemaker*.

sinus (si´nus): a cavity or hollow space within a body organ such as a bone.

skeletal muscle: a type of muscle tissue that is multinucleated, occurs in bundles, has crossbands of proteins, and contracts either in a voluntary or involuntary fashion.

small intestine: the portion of the GI tract between the stomach and the cecum, functions in absorption of food nutrients.

smooth muscle: a type of muscle tissue that is nonstriated, composed of fusiform, single-nucleated fibers, and contracts in an involuntary, rhythmic fashion within the walls of visceral organs.

somatic (so-mat´ik): pertaining to the nonvisceral parts of the body.

spermatic cord (sper´mat´ik): the structure of the male reproductive system composed of the ductus deferens, spermatic vessels, nerve, cremasteric muscle, and connective tissue.

spermatogenesis (sper´mah-to-jen´e-sis): the production of male sex gametes, or spermatozoa.

spermatozoon (sper´mah-to-zo´on): a sperm cell, or gamete.

sphincter (sfingk´ter): a circular muscle that constricts a body opening or the lumen of a tubular structure.

spinal cord (spi´nal): the portion of the central nervous system that extends from the brain stem through the vertebral canal.

spinal nerve: one of the thirty-one pairs of nerves that arise from the spinal cord.

spleen: a large, blood-filled organ located in the upper left of the abdomen and attached by the mesenteries to the stomach.

spongy bone (spun´je): a type of bone that contains many porous spaces; also called *cancellous bone*.

stomach: a pouchlike digestive organ between the esophagus and the duodenum.

submucosa (sub´mu-ko´sah): a layer of supportive connective tissue that underlies a mucous membrane.

superior vena cava (ve´nah ka´vah): a large systemic vein that collects blood from regions of the body superior to the heart and returns it to the right atrium.

surfactant (ser-fak´tant): a substance produced by the lungs that decreases the surface tension within the alveoli.

suture (su´chur): a type of fibrous joint articulating between bones of the skull.

sympathetic (sim´pah-thet´ik): pertaining to that part of the autonomic nervous system concerned with activities antagonistic to the parasympathetic.

synapse (sin´aps): a minute space between the axon terminal of a presynaptic neuron and a dendrite of a postsynaptic neuron.

synovial cavity (si-no´ve-al): a space between the two bones of a synovial joint, filled with synovial fluid.

system: a group of body organs that function together.

systole (sis´to-le): the muscular contraction of the ventricles of the heart during the cardiac cycle.

systolic pressure (sis´tol´ik): arterial blood pressure during the ventricular systolic phase of the cardiac cycle.

T

target organ: the specific body organ that a particular hormone affects.

tarsus (tahr´sus): pertaining to the ankle; the proximal portion of the foot that contains the seven tarsal bones.

tendo calcaneous (ten´do kal-ka´ne-us): the tendon that attaches the calf muscles to the calcaneous bone.

tendon (ten´dun): a band of dense regular connective tissue that attaches muscle to bone.

testis (tes´tis): the primary reproductive organ of a male, which produces spermatozoa and male sex hormones.

thoracic (tho-ras´ik): pertaining to the chest region.

thoracic duct: the major lymphatic vessel of the body, which drains lymph from the entire body except the upper right quadrant and returns it to the left subclavian vein.

thorax (tho´raks): the chest.

thymus gland (thi´mus): a bi-lobed lymphoid organ positioned in the upper mediastinum, posterior to the sternum and between the lungs.

tissue: an aggregation of similar cells and their binding intercellular substance, joined to perform a specific function.

tongue: a protrusible muscular organ on the floor of the oral cavity.

trachea (tra´ke-ah): the airway leading from the larynx to the bronchi; also called the *windpipe*.

tract: a bundle of nerve fibers within the central nervous system.

transverse colon (ko´lon): a portion of the large intestine that extends from right to left across the abdomen between the hepatic and splenic flexures.

tricuspid valve (tri-kus´pid): the heart valve between the right atrium and the right ventricle.

tympanic membrane (tim-pan´ik): the membranous eardrum positioned between the outer and middle ear; also called the *tympanum*, or the *ear drum*.

U

umbilical cord (um-bil´i-kal): a cordlike structure containing the umbilical arteries and vein, which connects the fetus with the placenta.

umbilicus (um-bil´i-kus): the site where the umbilical cord was attached to the fetus: also called the *navel*.

ureter (u-re´ter): a tube that transports urine from the kidney to the urinary bladder.

urethra (u-re´thrah): a tube that transports urine from the urinary bladder to the outside of the body.

urinary bladder (u´re-ner´e): a distensible sac in the pelvic cavity which stores urine.

uterine tube (u´ter-in): the tube through which the ovum is transported to the uterus and where fertilization takes place: also called the *oviduct* or *fallopian tube*.

uterus (u´ter-us): a hollow, muscular organ in which a fetus develops. It is located within the female pelvis between the urinary bladder and the rectum.

uvula (u´vu-lah): a fleshy, pendulous portion of the soft palate that blocks the nasopharynx during swallowing.

V

vagina (vah-ji´nah): a tubular organ that leads from the uterus to the vestibule of the female reproductive tract and receives the male penis during coitus.

vein: a blood vessel that conveys blood toward the heart.

ventral (ven´tral): toward the front surface of the body: also called *anterior*.

vestibular folds: the supporting folds of tissue for the vocal folds within the larynx.

vestibular window: a membrane-covered opening in the bony wall between the middle and inner ear, into which the footplate of the stapes fits; also called *oval window*.

viscera (vis´er-ah): the organs within the abdominal or thoracic cavities.

vitreous humor (vit´re-us hu´mor): the transparent gell that occupies the space between the lens and retina of the eye.

vocal folds: folds of the mucous membrane in the larynx that produce sound as they are pulled taut and vibrated; also called *vocal cords*.

vulva (vul´vah): the external genitalia of the female that surround the opening of the vagina; also called the *pudendum*.

Z

zygote (zi´gote): a fertilized egg cell formed by the union of a sperm and an ovum.

Index